Waste Not, Want Not-
A Lean Engineering
Approach To

American Warming

An Action Plan That
Everyone Can Agree On

By

Graham Stevens

Library of Congress Cataloguing-in-Publication Data

Stevens, Graham C., Waste Not, Want Not: A Lean Engineering Approach To American Warming: An Action Plan That Everyone Can Agree On

Includes bibliographical references.

ISBN-13: 978-1533347466
ISBN-10: 1533347468

1.   Science / Global Warming & Climate Change
2.   Business & Economics / Industries / Energy

Cover Image Photos: Wikimedia Commons, used under Creative Commons license: Drought.jpg by Tomas Castelazo, dry ground in the Sonoran Desert, Mexico; Forest-fire-CC.jpg by Cameron Strandberg from Rocky Mountain House, Alberta, Canada (DSC_7139); 800px-F5_tornado_Elie_Manitoba_2007 by Justin Hobson. See endnote[1] for further cover art attribution.

Second printing, May 2016

Dedication

To my family; and to my colleagues at Navigant Consulting Inc., who are one of the driving forces supporting U.S. economic growth.

"From the first day to this, sheer greed was the driving spirit of civilization."

— Friedrich Engels

"Greed, like the love of comfort, is a kind of fear."

— Cyril Connolly, *The Unquiet Grave*

"Fear and greed are potent motivators. When both of these forces push in the same direction, virtually no human being can resist."

— Andrew Weil

Dedication..3

What is Lean Engineering? 9

Why Bother...13
Guilt: We Need Fossil Fuel, But It's Wrecking the Planet............14
Depression: Cleaning Up Two Mount Everests' Worth 16
Fear: Global Warming Armageddon?................................23
    The Canaries in the Coal Mine......................... 30
Denial: Bah Humbug—Manmade Climate Change Hoax 32
The System is Rigged — But It's Not Your Fault....................39
Lean Engineering: Making Invisible Waste Visible 41

Changes To Activate the Market...........44
No Free Lunch...44
Eliminating the Other Free Lunch 48
Fines, Carbon Tax, or Cap & Trade? 50
    Carbon Tax Solutions............................... 50
    Cap and Trade 51
Fixing Other Market Distortions 55
    Subsidies 56
Energy Cost Myopia...61
Lean Engineering Market Activation Summary....................62

Expensive Fossil Fuels Reduction Plan 65
Fossil Fuel Usage and Emissions Drivers.......................66
    Population Growth 67
    The One Thing That Works......................... 70
    Historical Productivity Growth.................... 74
Economic Law of Pollution Applied............................81

Top U.S. Emissions Sources and Trends 81

Strategy #1: Improve Efficiency or Utilization of Resources ... 84

Strategy #2: Eliminate Pollution at the Source 88

Strategy #3: Capture Pollutants When Emitted 100

Strategy #4: Clean Up Afterward 108

Action Plan Summary 109

Short Term 109

Long Term 111

Comparison to Alternative Plans 113

How to Pay 118

Economics Background 118

Economic Growth Factors 119

Economic Brake Factors 119

Cluster Economics—Infrastructure for Growth 122

Globalization, and Free Trade 123

Finding Invisible Economic Waste 124

Government Waste 126

Extra Profit Based on Monopolistic Position 131

Rules/law Transparency 144

Rules/Law Consistency 145

Number of Laws/Rules/Regulations 145

Discretionary Position 146

Accountability 146

Market Distortion 146

Promoting Economic Growth 147

"A Penny Saved is a Penny for Jobs" 149

Lean Engineering Savings Summary 150

Timeframes: How to Accelerate Action............... **154**

How Much Time Do We Have? 154

How Not To Accelerate Action.................... 162

How To Accelerate Action 164

Insurance Plans..**167**

Conservation 167

Geo-engineering 169

Summary 172

What You Can Do Locally **173**

Appendix A: Glossary & Abbreviations....................**179**

Appendix B: Top 4 through 10 Emission Sources..................**183**

Appendix C: Improving Innovation **193**

Appendix D: Other Barriers to Action.....................**197**

Appendix E: Global Warming Feedbacks **204**

Appendix F: EROEI **207**

Appendix G: End Notes **211**

About the Author **250**

I

What is Lean Engineering?

"In God we trust; all others must bring data."
— W. Edwards Deming,
a founder of Lean Engineering

When we think about global warming, Americans are in denial.

Scientists say we need to cut our fossil fuel emissions by at least 80% to avoid global warming catastrophe. But fossil fuels drive our modern economy, so cutting fossil fuel emissions means voluntarily getting poorer, and going backward technologically.

So we refuse to go on a "mitigation" low-carbon diet because it's too expensive, even though "adaption" will be more expensive in the distant future. So we keep kicking the can down the road, hoping that technology will save us.

And we're right—technology might save us. But right now, with our current societal economic structures, it can't. Technology needs a little help from us, from capitalist greed, and from lean engineering principles used in manufacturing.

Lean engineering is a systematic method for eliminating waste within a manufacturing environment. Modern methods started with Eli Whitney's cotton gin and his use of interchangeable parts, and continued with Frederick Taylor's Scientific Management and Ford's assembly line for the Model T. The principle behind Ford's assembly line won World War II by overwhelming the Axis with massive quantities of material[2]. After the war, the Japanese improved the assembly line and made it more flexible at Toyota. Their improvements spread back to the U.S. in the 1980s and 1990s, under the monikers "Just-in-Time" manufacturing and "Lean Engineering." Lean engineering continued to evolve, encompassing a large variety of continuous improvement systems, including "5S", Six Sigma, Total Quality Management, Kanban, etc., and is now spreading to non-manufacturing arenas such as hospitals, retail, government, non-profits, and other service industries.

At Toyota, a management style arose that focused on having groups of employees systematically work on eliminating production waste. From a manufacturing perspective, anything that doesn't produce or add value is wasted motion, and should be eliminated. The ideal is a manufacturing process flow with zero waste. The key to seeking this ideal is a constant and continuous focus on identifying and eliminating losses. The Japanese separated manufacturing losses and wastes into three categories:

1) **Muda (waste)**
    a. **Transportation**. Moving stuff around does not add value.
    b. **Inventory.** Raw materials, work-in-process, or finished products[3] not being actively processed are wasted capital not producing income.
    c. **Motion**. Extra motions at work stations, including accidents or repetitive stress injuries, do not add value.
    d. **Waiting.** When work is not in transport or being processed, it is waiting and not adding value. In traditional manufacturing, in-process materials spend most of their time waiting.

e. **Over-production.** This occurs when more product is produced than required, or sooner than required, by a customer.

f. **Over-processing.** This occurs when more work is done than is required by a customer, including higher precision, complexity, and quality.

g. **Defects.** When errors occur, extra costs are incurred to inspect, rework, re-schedule, etc.

h. **Talent.** Failing to use the skills and knowledge of all employees.

i. **Resources.** Wasting energy, water, or other resources used to make a product

j. **Byproducts.** Not making use of the byproducts of your processes.

2) **Muri (overburden)**

a. **Poor organization.** This can include lack of training, wrong tools, unclear instructions, etc.

b. **Poor design.** Lack of standardization.

c. **Resources are stressed.** When unreasonable expectations occur relative to people or equipment, breakdowns occur.

3) **Mura (unevenness)**

a. **Fluctuating Demand.** Variations in demand or schedules are handled by creating a buffer inventory—but this inventory is wasteful. Solutions include reducing the total length of a supply chain, reducing delivery times between links, and creating transparency between links in a supply chain.

b. **Fluctuating Product Mix, or Production Methods.** These give rise to lower than necessary production volumes, increasing costs.

c. **Fluctuating Process Conditions.** When process conditions change, this introduces variability in the product. If too large, these conditions give rise to defects.

In sum, one can think of Lean Engineering as the modern extension of the assembly line and our ability to efficiently produce lots of stuff. We, as Americans (together with our outsourcing partners in China and elsewhere), are already very, very good at it.

But how will American excellence at producing lots of stuff solve global warming without harming our economy?

Let's find out.

II

Why Bother

"In utilizing and conserving the natural resources of the Nation, the one characteristic more essential than any other is foresight.... The conservation of our natural resources and their proper use constitute the fundamental problem which underlies almost every other problem of our national life"

— Teddy Roosevelt,
26th President of the United States[4]

Lean Engineering is all about identifying, measuring, and ultimately eliminating waste that is otherwise invisible or not readily apparent. But first, there are invisible factors and distortions of the truth that affect how we feel about global warming. Before we can take a hard-nosed lean engineering approach to develop solutions, we need to clear out some invisible emotional baggage.

While searching for global warming solutions from all points of view, we find four common and paralyzing emotional reactions:

(1) Guilt ("My emissions are wrecking the planet")

(2) Depression ("It's too big to deal with, it's hopeless")

(3) Fear ("We're all gonna die!")

(4) Denial ("It's not real").

Let's briefly consider each of these, so we can dispassionately look at affordable solutions in the next chapter.

Guilt: We Need Fossil Fuel, But It's Wrecking the Planet

Most[i] scientists that study climate and weather tell us that our exponentially increasing fossil fuel CO_2 emissions appear dangerous and eventually may lead to a 60 foot rise in sea levels. But the higher a society's wealth, the higher fossil fuel emissions[ii] are:

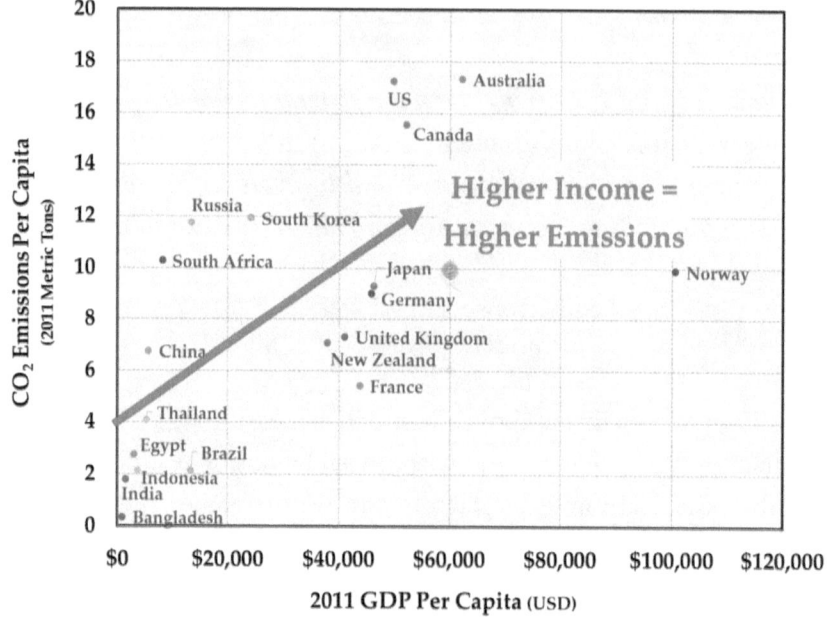

Figure 1. GDP vs. CO_2 Emissions for Select Countries[5]

---

[i] Let's avoid the semantics of the word "consensus" or 97%, and agree that most equals "greater than a majority"; see, for example, Ian Tuttle "The 97% Solution", National Review, Oct 2015. 32,000 signatures from the Petition Project, vs. ~7 million STEM workers in the US (www.bls.gov), is much less than 50%.

[ii] When burned, the carbon in gasoline, coal and natural gas produce heat and CO_2 gas which is emitted into the atmosphere. The terms "CO_2 emissions", "fossil fuel emissions", and "carbon emissions" are therefore used interchangeably throughout the book.

So we feel guilty, because we refuse science's repeated call to reduce consumption and emissions, because that would lower economic growth and make us poorer.

Modern economic growth is wholly dependent on fossil fuels. Economic growth and lean engineering first began with the discovery of fire about a million years ago. Cooking our food allows us to spend only 5% of our time eating and chewing, rather than wasting about half of our time like chimpanzees do[6].

After fire, human's use of tools was relatively static until the Industrial Revolution, when coal freed most of us from the burden of growing food. Horses reduced manual labor, and then larger and more powerful machines replaced horses. Our entire economy now uses inexpensive energy that allows individuals to accomplish exponentially larger amounts of work.

Since the industrial revolution, farmers decreased from over 90% of the population to less than 1% today, the number of workers in the extraction and energy industries shrank to less than 4%, and more than 95% of the economy involves jobs that use this extracted energy and resources. Labor productivity and wealth grew exponentially. While the U.S. population has tripled since 1914, our economic output grew over 450-fold over the last century, powered by fossil fuels and technological innovation. Railroads, steel, industrial factories, electricity, cars, airplanes, air conditioning, global telecommunications, computers and the Internet, economic growth— none of these would have been possible without exponentially increasing flows of fossil fuel energy that was itself inexpensive[7].

In addition to global warming threatening to be too expensive, some Americans may also feel guilty because we are relative carbon emissions spendthrifts compared to everyone else:

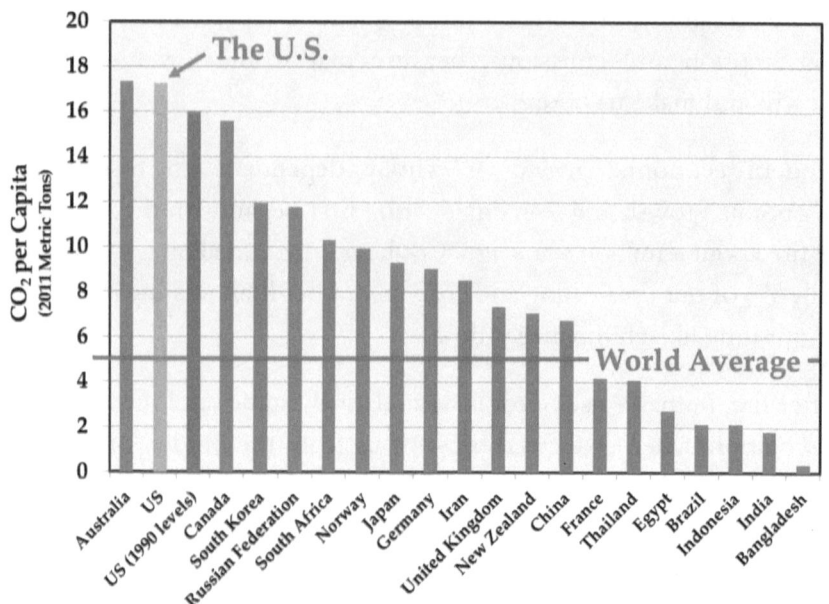

Figure 2. 2011 Emissions Per Capita for Select Countries[8]

As Figure 2 shows, the U.S. annually emits 17 metric tons of CO_2 per capita, one of the highest overall. Our per capita emissions are triple the worldwide average of 5, and roughly double what those with comparable lifestyles in Europe and Japan emit.

The cure for our guilt is to recognize that we can have economic growth, and reduce fossil fuel emissions at the same time[9].

Depression: Cleaning Up Two Mount Everests' Worth

Global warming is overwhelming and depressing because our pollution mountain is so large. To truly clean up our mess, if the scientists are right, we must attend to 375,000 million metric tons[iii] of invisible CO_2 that we have dumped since the beginning of the Industrial Revolution:

---

iii 1 tonne = 1,000 kg (a metric ton), and equals ~ 1.1 U.S. tons.

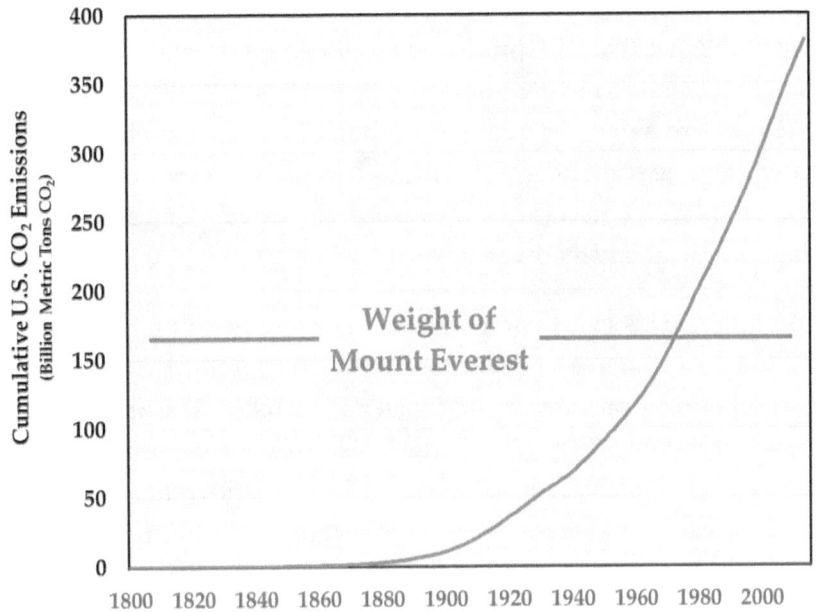

Figure 3. America's CO₂ Emissions Cleanup Mountain

To put this number into perspective, this weighs at least as much as two Mount Everests. And how can I, as one American, do anything about these mountains?

The answer is ... exactly like the previous times we had to clean up the environment:

The first step is to acknowledge the problem. Virtually all pollutants are not thought of as pollutants until their effects are proven to be harmful. Cigarette smoke, DDT and other pesticides, mercury and other heavy metals, ozone and smog, fertilizer run-off, sulfur dioxide /acid rain, etc., are all examples of this phenomenon. We used these substances until the harmful effects became apparent. After outcry and denial, environmental regulations were implemented to reduce the harm.

Carbon dioxide is no different, as evidenced by recent Supreme Court decisions that have upheld the EPA's authority to regulate it as a pollutant[10], despite the recent stay of the Clean Power Plan[11]. However, a key problem is that CO_2 production is ubiquitous in our lives— fossil fuels are integral to our economy. So the economic stakes and costs of abatement are higher. In addition, carbon dioxide is invisible, and it's harder to prove something is harmful when you can't see, touch, smell, hear, or taste it. Its effects take place at a much larger and more complex scale than any other substance.

Carbon dioxide is everywhere in our air— how can it be harmful? In fact, more of it increases plant respiration and productivity by up to 40%, depending on nutrient and water availability. This is called the CO_2 fertilization effect[12], a negative feedback mechanism we have observed[13]. But plant growth is limited by nutrients other than CO_2, including water. Higher temperature modeling of the future shows less water and deserts growing, which could reduce the growth of plants more than the fertilization effect would increase it. As a result, global food growth is projected to suffer at higher temperatures. Insects and disease are also projected to increase at higher temperatures. But, because there are so many variables, no one really knows.

We do know, though, that virtually everything on the planet is a potential pollutant, depending on context and dosage. Water, if you drink too much, can kill you[14] —6 liters is enough to kill a 165 pound person[15]. Play sand in California (and everywhere else) contains silica, a carcinogen. Oxygen, at higher pressures or concentrations, can also kill[16].

So was the Supreme Court correct to name CO_2 a pollutant? A pollutant is defined as a substance that has harmful or poisonous effects. While many chemical poisons are obviously harmful, other substances can be harmful while not being poisonous. For example, plastics are not poisonous in small quantities, but are considered pollutants when they litter the land or oceans; phosphate fertilizers aren't poisonous either, but can cause harmful algae blooms in waterways.

18

Aside from its heat trapping properties, CO_2 in the atmosphere finds its way into the ocean, turning the ocean more acidic. It is therefore harmful, and a pollutant.

When pollutants are not recognized as pollutants, and therefore no fines or other penalties are applied for their use, the market cannot fundamentally value them. This is known as an "externality" in economic circles; participants make market decisions without considering the harm done to the commons[17]. During this period which we are in now, the market subsidizes CO_2 pollution by not taking into consideration its harmful effects. This distortion leads to higher emission levels than might otherwise occur. With fossil fuel carbon dioxide pollution, there are no economical technically feasible replacements in many spheres, so we are stuck with CO_2 pollution as a consequence of our modern lifestyle and economy.

The lack of both carbon-based fuel alternatives and financial incentives to seek them inhibit private sector investments in low-carbon technologies and R&D. This is because inventor royalty rates are relatively low compared to what society gains from the inventions as a whole.

For example, if I invent (or invest in) a device to increase automobile efficiency, I will get about 5% of the price of that device, while consumers will save hundreds of dollars annually on fuel. With a zero carbon price, "projects to develop promising low-carbon technologies will not get to the boardroom of a profit-oriented company," and this doubly discourages profit-oriented R&D on low-carbon technologies[18]. As currently structured, our capitalist system literally can't solve the problem, because inventors and their investors won't get paid.

Because fossil fuels are so integral to our modern economy, any change in how we ignore CO_2 emissions will cost the fossil fuel industry a lot of money. Industries that use a lot of fossil fuels (such as airlines, automobiles, electric utilities, etc.) would also be disproportionately affected. There are winners and losers with any change in the status quo, and the fossil fuel sector will therefore fight any substantive change. As many of these industries are also central to the economy, any change could potentially slow economic growth as a whole[iv].

In response, the fossil fuel industry is adopting the same tactic previously used by the smoking and chemical industries: Introduce doubt regarding (a) whether there is a problem, (b) how big it is, and (c) what to do about it. The "Merchants of Doubt"[19] are still as effective as they were for smoking and chemicals, but the stakes are much larger. For example, economist William Nordhaus points out that tobacco industry sales are about 30 billion dollars, compared to 1 trillion dollars of energy sales in the U.S.[20]

The doubt tactic succeeds because denial is a common psychological reaction to disasters[21] and change, especially personal loss[22]. When scientists say we need to reduce fossil fuel usage, and our wealth and exponential economic growth depend wholly on fossil fuels, America's first reaction is denial. Add in overwhelming complexity and the scale of the problem, and climate change denial is widespread among Americans[23], with 20% of Americans not believing in global warming[24]. Even among the 80% of us who do believe in global warming, our attitudes reflect different shades of denial[25].

With regard to environmental pollution, humans follow a single pattern.

The worldwide historical environmental record shows, over and over again, that humans have a history of ignoring and denying the pollution we create, making a mess, and cleaning up afterward. This was true in London, when the "Great Smog" of 1952 killed 4,000 people, and precipitated the U.K. Clean Air Act of 1956; it was true of oil spills

---

iv But this is a smokescreen. Other industries would likely grow to compensate for a slowdown in fossil fuel growth, offsetting fossil-fuel sector related losses.

(Santa Barbara oil spill of 1969), factory pollution (the Cuyahoga river fire of 1969), and pesticide use (aka Rachel Carson's "Silent Spring" [26]) in the U.S., which led to the creation of the Environmental Protection Agency; and it is true in China today, with a new push to clean up the environment.

It is true for global warming CO_2 pollution. But also unfortunate.

The "clean up afterward" strategy may not be possible (i.e. we may reach tipping points faster than we can change), and this strategy invariably is the most expensive solution. At $500-$1,000 per metric ton or higher for direct air capture solutions, cleaning up afterward also looks like it will be prohibitively expensive— greater than 25% of U.S. gross domestic product annually for 75 years. Cleaning up afterward, which is what humans have always done, really *will* kill our economy. Because of this huge cost, it's no wonder we deny its necessity, and pray that technology can rescue us.

Are there less expensive solutions? Yes, but they aren't technical. If America gets started cleaning up our more than 375,000 million metric ton pollution mountain one step at a time, and applies our pollution control knowledge to CO_2, we can solve the problem inexpensively. But, as we have seen for all environmental cleanup jobs[27], it takes years and decades to clean up after ourselves.

America excels at cost-effective atmospheric pollution control, and several lean engineering strategies have emerged in the last century. In order of increasing cost, they are:

Table 1. The Economic Law of Pollution

Strategy	Examples	Cost
Improve efficiency or utilization of emitting resource	Improving electricity plant or end-use efficiency, low flush toilets, detecting and fixing water leaks, recycling	Lowest
Eliminate pollution at the source by substitution	Substitution of natural gas for coal to produce electricity	Less
Capture/convert pollutants as they are being emitted	Car tailpipe catalytic converters, factory stack scrubbers	More
Clean up afterward	Treating polluted water, Superfund cleanup, etc.	Highest

Per the table title, let us call this empirical collection of our experience reducing pollution the "Economic Law of Pollution." It applies Benjamin Franklin's adage "a penny saved is a penny got"[28] and lean engineering (where reducing waste is the least expensive option) to pollution.

These control strategies, and the technological innovation they have spurred, have reduced emissions in the United States. Between 1970 and 2010, the EPA substantially lowered emissions of eight pollutants: lead (reduced 97%), mercury (reduced more than 45%), benzene (more than 64%), carbon monoxide (68%), ozone (28%), nitrogen dioxide (58%), sulfur dioxide (82%), and VOC emissions (55%). The entire country now generally attains national EPA air quality standards, while the gross domestic product has increased 130%, vehicle miles traveled have increased 94%, and population growth has increased 37% during these last four decades. Costs have been minimal, on the order of less than .5% GDP.

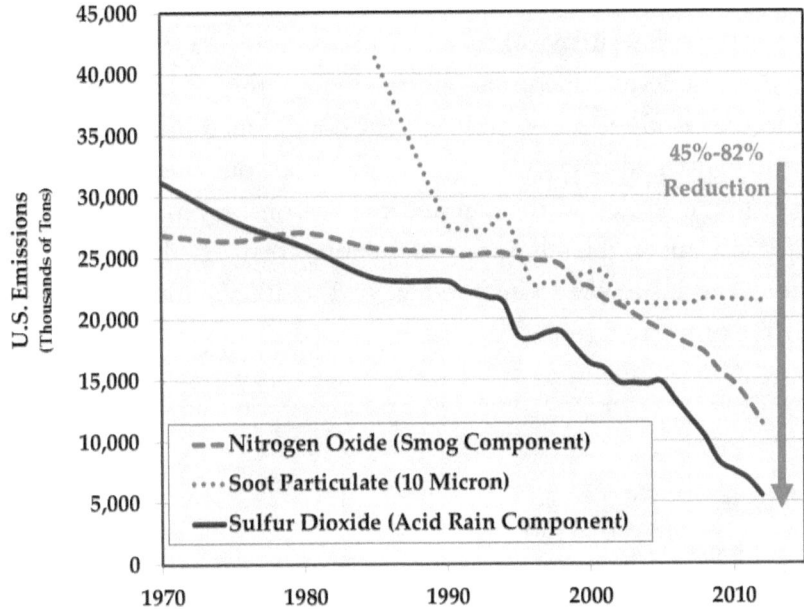

Figure 4. U.S National Emissions Reductions Since 1970[29]

So we can have our cake and eat it, too—if we apply lean engineering to the problem. .

Fear: Global Warming Armageddon?

We see the news reports. The permafrost is melting[30]. Methane hydrates are releasing bubbles of methane in the East Siberian Sea[31]. The Western Antarctic Ice Sheet is melting from below and is unstable[32]. The Totten Glacier of Eastern Antarctica is being eroded from below[33]. "It's crazy to think that 2 degrees Celsius is a safe limit."[34] The consistent message from global warming science is that if we don't do something yesterday, sea levels will rise catastrophically.

But will catastrophe occur? What does the data say about the real costs of global warming to the U.S.? Is it so bad?

In the short term, no.

When people think about global warming or climate change, we associate it with weird weather—heat waves, rising sea levels, flooding, etc. But modern Americans mostly ignore the weather and our connection to the land —we live in air conditioned boxes, and have electric lights and devices so we can go about our business without regard to the cycles of the sun and the seasons. Even farmers—the profession most directly affected by the weather—have financial options and insurance to help reduce bad weather's impact. And less than 1% of us are farmers[35] anymore.

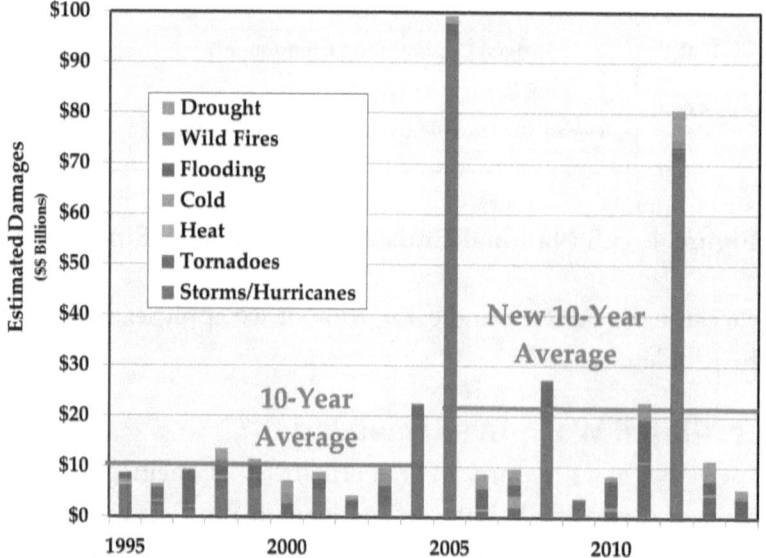

Figure 5. Annual U.S. Weather Related Damage[36]

The level of damage in the U.S. from "hundred-year and thousand-year" large storms—from Hurricane Sandy to Hurricane Katrina—has doubled on average in the last decade. Hurricane Katrina cost over $108 billion in federal support, over time, and with its wider swath, the final tally from Sandy might reach this amount[37]. Current estimates are over $50 billion[38], which is .3% of the $17.4 trillion U.S. GDP for 2014—so while pricey, this is chump change relative to our overall economy.

So, if we have conquered the weather, and global warming is primarily weather-related, why should we care about global warming? The storm damage levels shown in Figure 5 are increasing, sure, but are still manageable relative to the large U.S. economy.

Killing U.S. citizens gets our attention and causes America to mobilize for war—but it is hard to directly attribute weather-related deaths to global warming. Over 1,200 people died in Hurricane Katrina; and 147 died in Hurricane Sandy —but we can't tell whether these storms, and their severity, are due to natural fluctuations, or to global warming.

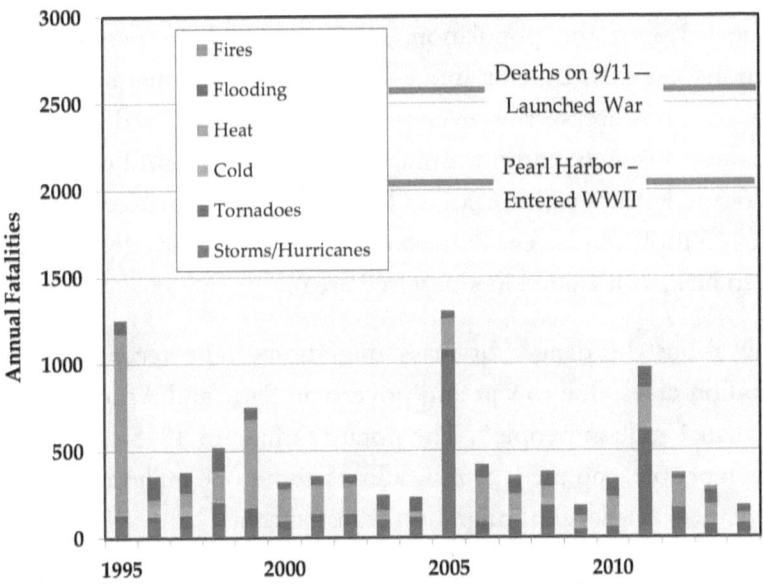

Figure 6. Severe Weather Related Fatalities in the U.S.[39]

Overall, the frequency and severity of storms in the U.S. may have increased over the last decade. But the sample size, a few super-storms, is too small to make solid conclusions that link global warming with storm damage. Even when one expands the scope to examine storms worldwide rather than just in the U.S., researchers have found that storm damages have been going up, but that increased exposure and the value of capital at risk is the real driver, not changes to the climate[40].

Humans like to live in expensive buildings on the coasts where storm damage is most likely to occur.

The U.S. is therefore relatively insulated from storm damage and weather damage because of our wealth. However, the rest of the world will be hurt more by land loss, storm damage, and costs, especially the poor nations where about 5/6ths of the worldwide population of over 7 billion people live. Just one example is the Philippines, where 6,241 people lost their lives in supersized cyclone Haiyan in 2013[41] (exceeding the fatalities of all storms in the U.S. for the past 5 years). Haiyan displaced 4 million people, at a cost of $809 million. Worldwide about 1-3% of the population, or 70-200 million people, live in locations less than 1 meter above sea level, many of them in countries with few resources. But even there, humans are reducing storm casualties through early warnings and preparation. For example, Hurricane Patricia was the largest hurricane to land in North America, with 165 mph winds. Yet Patricia caused no deaths in Mexico (in part due to luck, as it landed in a deserted area).

What about the danger of mass migrations? The recent European migration crisis, due to war and poverty in Syria and Africa, numbers less than 1 million people[42]. The potato famine of 1845 displaced 1.7 million people, and the U.S. now admits about one million immigrants per year, so this level of migration is manageable.

As sea levels rise a foot over the next 85 years as predicted by the Intergovernmental Panel on Climate Change (IPCC), migrations may grow, especially in Asia and Africa. In the U.S., an estimated 4 million out of 318 million people live within a few feet of high tide[43]. But mass migrations will likely be spread out over many years as sea levels rise slowly (the last few decades of sea level rise in the U.S. have been about 1/8 inch a year), and so should therefore be relatively manageable.

Even the potential "death by a thousand cuts" effect of global warming—the smaller storms that cause damage nationwide, the costs to mitigate rising sea levels, higher cooling costs, and the costs to address drought in some areas and flooding in others—are manageable compared to the size of our economy.

But the long term is another story. If we continue our 250-year history of invisible CO_2 dumping, we will face an enormous cleanup bill.

How long is "long term"? The most serious impact of global warming —large increases in sea level that will engulf Florida and our Eastern seaboard cities— will likely take hundreds of years to occur, for two reasons:

(1)   The ocean's heat capacity and thermal inertia is high, so heat takes a long time to be absorbed—similar to water boiling on your stove. In scientists' search for the missing heat causing the global warming surface temperature pause from 1998-2013 (also called the global warming hiatus), the ocean's temperature increased only .1C in the last decade[44]. Earth is a water covered planet, so its responses to heat are muted because it takes an enormous amount of energy to change water's temperature. This climate lag effect slows down the climate's responses to forces that can shift Earth's energy balance.

(2)   There are massive bodies of ice that need to melt. The GRACE satellite project measured average ice losses from the Greenland ice sheet of about 164 gigatons (Gt) between 2003 and 2013[45]. With the Greenland ice sheet's weight estimated at 2.8 million giga-tons[46], it will take over 16,000 years to melt at current rates. Even if we use the maximum ice loss observed in any one year over the last decade, 474 Gt, this still means it will still take 5,800 years. The Intergovernmental Panel on Climate Change's 5th Assessment Report (AR5) value was 215, implying over 12,000 years.

But in 2015, James Hansen and other researchers projected Greenland melting in 50-100 years, not 12,000[47]. How? Hansen posits non-linear feedback mechanisms that would *exponentially* increase the rate of melting Antarctic ice shelves from warm water below, rather than a constant rate of melting. No one knows whether such a hyper-positive feedback mechanism is relevant, but we know that (a) the IPCC has generally been late and conservative in many of non-temperature predictions (Artic sea ice, sea level rise, etc.[48]) and (b) the IPCC (and Hansen) incorrectly did not predict the temperature hiatus, and underestimated negative feedback mechanisms.

The true answer most likely lies somewhere in between, leading me to the conclusion that hundreds of years is correct. So we have time. Nevertheless, this does not mean we can ignore long-term global warming impacts.

Unfortunately, we cannot rely on climate models to assess, predict, or warn us regarding tipping points[v] over the long term because we don't know enough to predict these. Climate models have been consistently wrong[49], did not predict the hiatus[vi] where surface temperatures have been relatively flat for a decade, and/or have underestimated negative feedbacks[50], the impact of natural variability, climate lag[51], or latency[52]. This is not surprising, because there is a world of variables out there. Modelling climate change mathematically is daunting, albeit the science is improving slowly. We are stuck with approximations which are inaccurate to some degree. In addition, required initial conditions and historical conditions are very hard to measure accurately worldwide.

---

[v] A "tipping point" occurs when a number of small changes add up to a big one. With regards to climate change, it is an abrupt change that will likely not be reversible in our lifetime. Examples may be: enough of Greenland ice melts to trigger "permanent" ocean circulation changes.

[vi] And the controversy regarding its existence. The recent NOAA paper shifting inaccurate bucket ocean temperature historical datasets slightly colder so that the hiatus was erased seems just a bit too convenient ... rather than relying on such data, we need to admit the historical error bars are too large. This does not change the fact that our Earth is warming relatively quickly.

So what we have is a "garbage in-garbage out" data situation to some extent, especially regarding aerosols, clouds, and water phase transitions on a planet whose surface is 75% water.

Climate science and modeling has made great strides, but our knowledge of tipping points is still in its infancy. The Earth's sea level has been homeostatic[vii] during the rise of human civilization over the last 8,000 years, despite our being in a climatic cooling period relative to Milankovitch solar forcings[viii] that have triggered ice ages and retreats in the past:

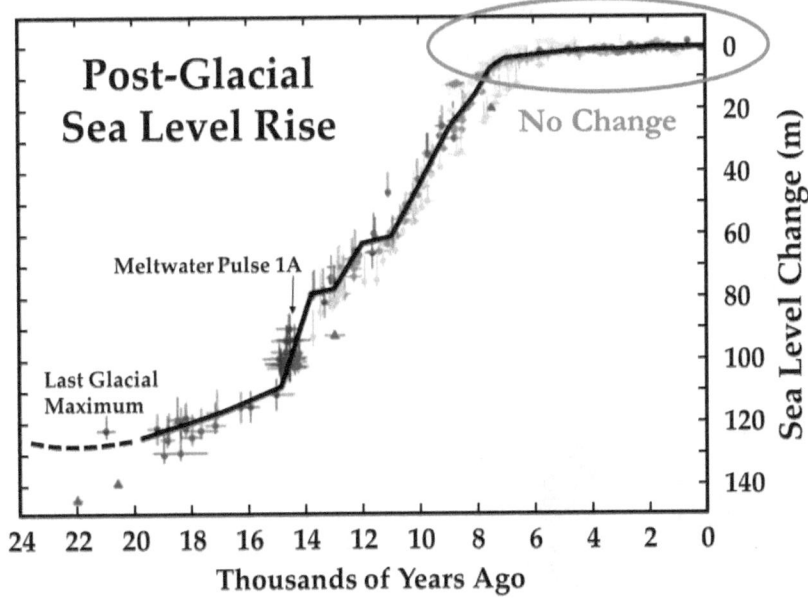

Figure 7. 8000 Years of Sea Level Stability[53]

[vii] A homeostatic system resists changes and stays constant. An example is human's internal temperature staying at 98.6F, no matter what the temperature is outside.
[viii] Changes in the tilt of the Earth's axis, precession, and/or eccentricity change how much solar radiation the Earth's poles receive, triggering ice ages and retreats over many thousands of years. These types of changes that can affect Earth's energy balance are called Milankovitch "solar forcings," named for their discoverer.

Despite this homeostasis, the U.S. has recently gotten some very weird weather. The jet stream shifted in winter 2014 so most of the snow got dumped on the northeastern U.S., while the West was unseasonably warm. As I sit here, in late October in Moscow, Idaho, raindrops fall outside my window instead of the usual snow. We can feel that the seasons have shifted[54].

The Canaries in the Coal Mine

There are many signs that the Earth's climate has shifted beyond natural variations. The Arctic Sea ice is receding[55] enough to open the Northwest Passage[56] to commercial traffic for the first time in human memory. Glaciers are melting worldwide[57]. The Earth's poles are more sensitive to global warming changes than areas near the equator, and recent changes in Arctic ice cover are shown in Figure 8. 2015 ice cover is 25% below the average minimum from 1979-2015.

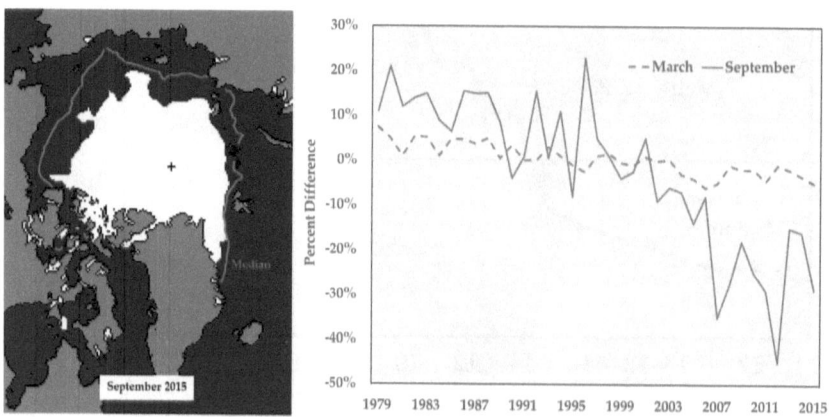

Figure 8. Sea Ice Trends at the North Pole[58]

Spring is coming early, and fall is arriving late[59]. Sea levels are rising[60], and the ocean is becoming more acidic[61]. Ecosystems are shifting toward the poles[62]. Storms are causing more damage (albeit perhaps less frequently), wildfires and heat waves have increased, and the weather is getting more erratic (both too hot and too cold at times)[63].

However, the problem remains that we cannot effectively model precisely how hot it will get and when or where. We can't predict total impacts and potential tipping points. In addition, the large latency in Earth's climate system means that yesterday's emissions will continue to affect us for centuries, even if we ceased all CO_2 emissions today (which isn't likely).

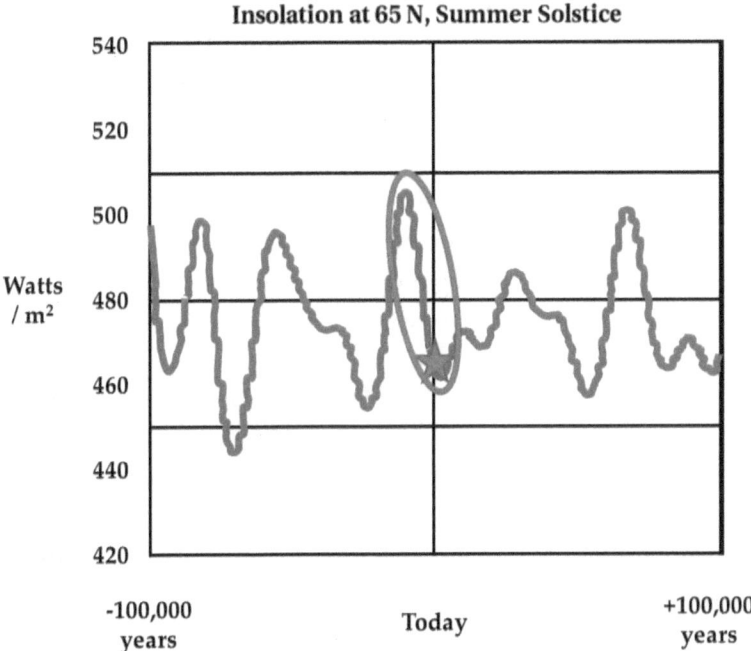

Figure 9. Milankovitch Solar Insolation[64]

This shift from normal weather patterns has occurred in a time when Earth should be getting colder long term, not warmer. Figure 9 shows the Milankovitch solar cycles[65] that climatologists have found to be a primary factor in how ice ages come and go. As shown by the ellipse, we are in a solar cooling phase at present.

This graph offers proof that global warming is man-made. Polar ice melting, seasons changing, and ocean warming by .1 degree C all take an enormous amount of energy, and science has found no reasonable explanations beyond rising CO_2 and other greenhouse gas levels that are much larger than natural oscillations[66]. It is therefore clear that mankind has changed the planet's climate, shifting it from a cooling regime to a heating regime, as shown by the canaries in the coal mine above (and admitted by the U.S. Senate).

We need to reverse this change. While we may have centuries, making these changes will take a long time, and therefore we need to do what we can do cheaply to get back down to pre-Industrial Revolution CO_2 levels of 280-350 parts per million (ppm) as soon as possible.

Denial: Bah Humbug—Manmade Climate Change Hoax

Let's presume for a moment that global warming deniers are correct, that global warming is unconnected to manmade CO_2 levels. Even if this were so, we still need to begin decarbonizing our economy. Let's examine the full scope of the problem with fossil fuels.

Remember peak oil, M. King Hubbert's theory that worldwide oil production peaked in 1970? It is fully debunked now as the U.S. is awash with new oil supplies from fracking technology, so much oil that the U.S. has now become the top oil producing nation in the world, beating Saudi Arabia's 9.6 million barrels of production, as Figure 10 below shows.

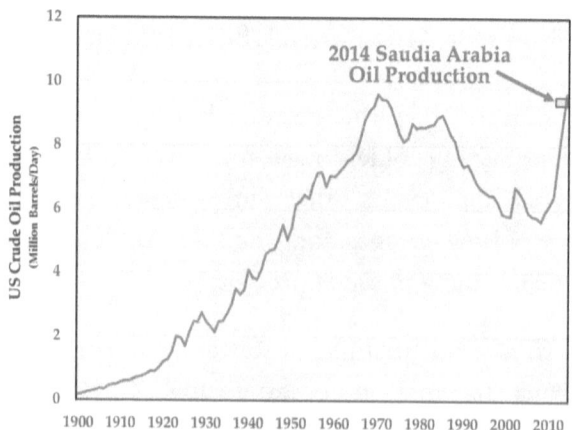

Figure 10. U.S. Oil Production[67]

Well, not so fast. Oil prices have come down from $100 a barrel highs, but are still over $30 a barrel. Many conventional sources of oil, with extraction costs of less than $1 a barrel, are exhausted; shale oil well break-even costs are over $25 a barrel, not counting financing. The extra cost of sand, water, fracking and re-fracking, and the geology of tight oil formations, will form a floor oil price. While we have new supply, the value of each worldwide barrel of "black gold" is lower because it costs more to dig up and process.

Ever since the 1970s, whenever oil prices have spiked, the U.S. economy has entered a recession. Current high oil prices therefore continue to drag down our economic recovery[ix], despite the welcome 2015 dip to less than $35 a barrel.

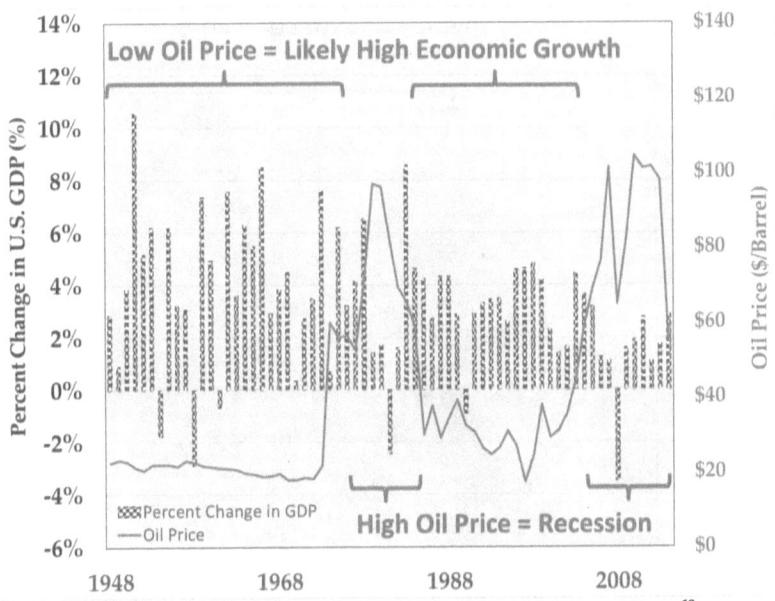

Figure 11. U.S. Recessions and U.S. Oil Prices[68]

It is still unknown how long our shale oil bonanza will last, as countervailing factors include increasing improvements to technology,

---

[ix] Note that there are many causes of recessions, especially financial industry breakdowns.

oil prices that impact investment, higher well depletion rates, total wells drilled, costs and availability of water, and environmental regulations. But shale oil reserves were always at the bottom of the barrel and hard to access, so supplies will last somewhat less than the 70 years of conventional oil production the U.S. enjoyed between 1900 and 1970. Figure 10's peak in the 1970s shows a clear warning of what is to come.

While we may not know the exact timing of when oil will peak again, we know that it will peak, whether in 2050-2060 as the Energy Information Administration (EIA) estimates, or otherwise, because oil—or at least economically recovered oil accessible with current technology —is a finite resource.

Similarly, coal production in the U.S. peaked in 2005-2008 because reduced mine productivity increased extraction costs.

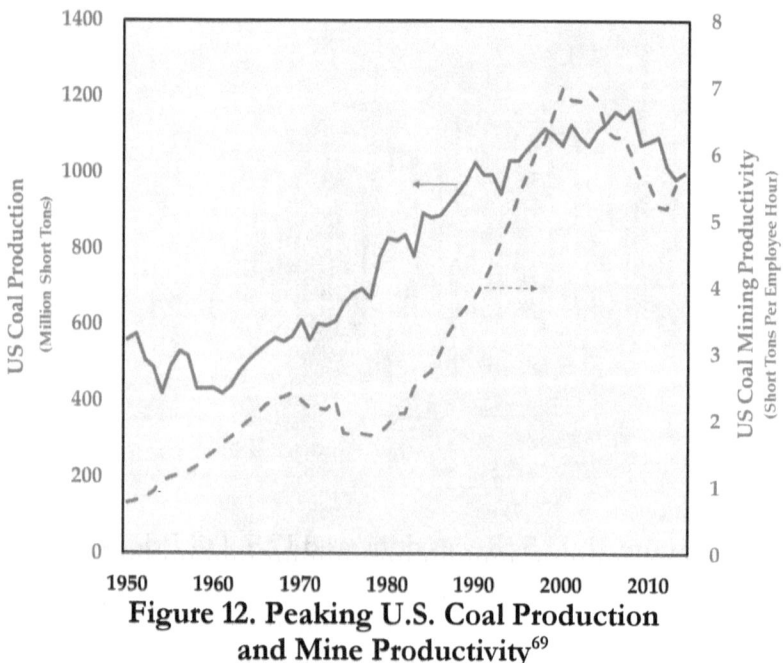

Figure 12. Peaking U.S. Coal Production and Mine Productivity[69]

Coal's cost competitiveness with natural gas for electricity generation has plummeted, and this is responsible for electric utilities recently switching to natural gas. But this switch has not been caused only by reduced natural gas prices due to fracking, or by pressures from potential EPA regulation changes. In the U.S., coal prices have increased over 64% since 2000 due to lower mine productivity. This has led to about 12% average inflation-adjusted increases in U.S. electricity prices since 2000[70]. The large 64% increase's impact on electricity prices is diluted by two factors: only a small proportion of electricity price is fuel, and non-coal sources supply about 60% of our power.

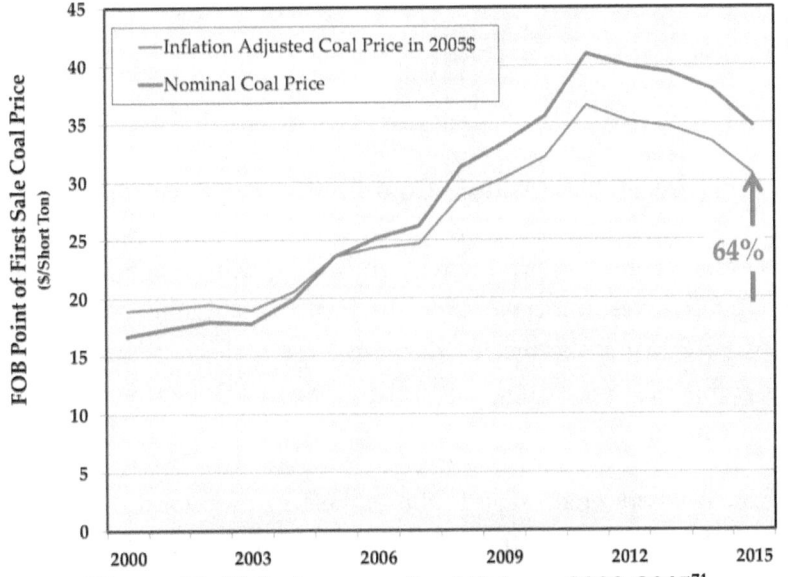

Figure 13. U.S. Average Coal Prices, 2000-2015[71]

Decreasing mine productivity occurs because (a) more overburden must be removed to get to the coal, and (b) coal deposit quality is deteriorating. Figure 14 on the next page shows how the highest quality grade, anthracite, was exhausted long ago; bituminous production peaked in the 1990s, and sub-bituminous production peaked in 2006. Lignite, the lowest grade, is only slightly more energy dense than renewable hardwood.

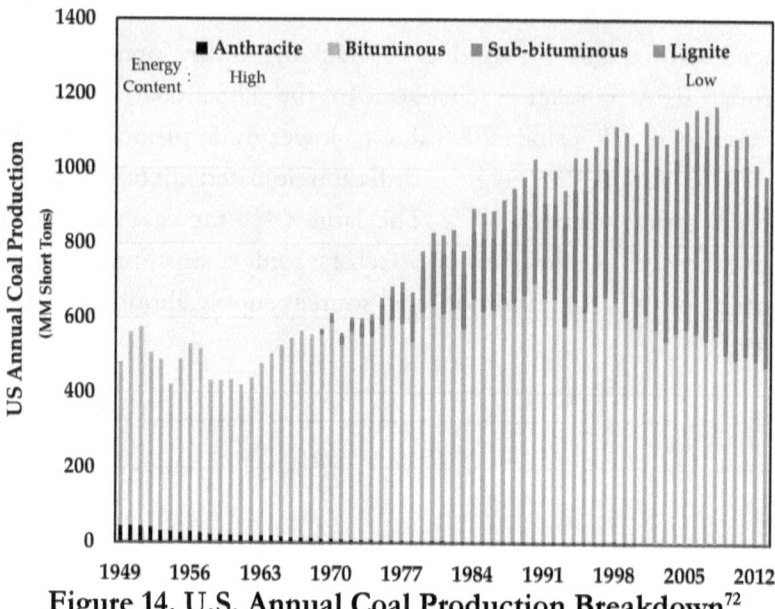

Figure 14. U.S. Annual Coal Production Breakdown[72]

This decrease in mine productivity and coal quality occurs throughout the United States, but is worst on the East Coast. Large operations on the West Coast make money through large scale, but barely; overall coal companies are struggling[73] relative to low-priced natural gas or going bankrupt[74].

In China, where coal usage is the highest in the world, the Chinese imported coal from Australia and other nations[75] in 2009; this would not have occurred if rising demand did not exceed their supply. Similar to U.S. trends, the price of coal supplied to China has doubled since 2000, from about $40 a metric ton to $80, or higher[76].

Increased oil and coal extraction costs (and natural gas, as fracking is not free), mean that we spend more of our GDP on energy, leaving fewer dollars to be spent on anything else, including consumer goods and climate change adaption or mitigation. As Figure 15 shows, our energy extraction costs have approximately doubled since 2000, from

2% to 4% of GDP. This is equivalent to about $350 billion more annual spending on fossil fuels, or $1,100 per person[77].

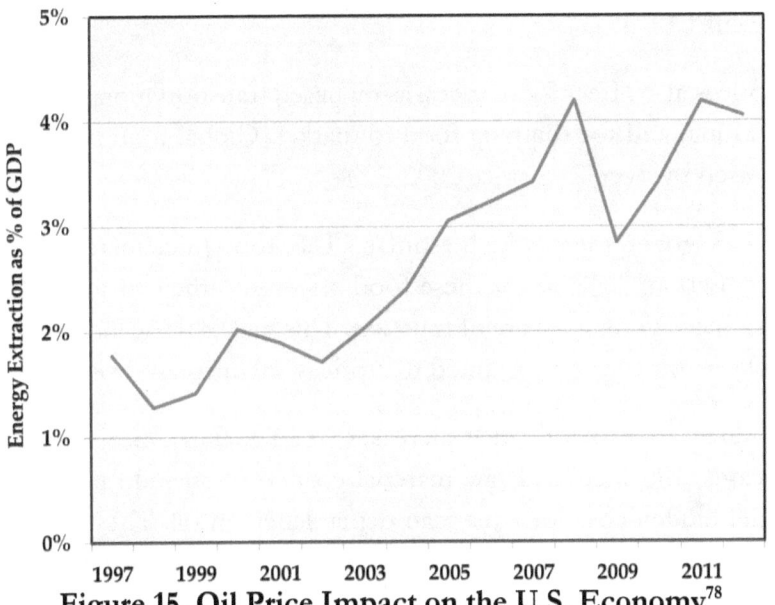

Figure 15. Oil Price Impact on the U.S. Economy[78]

The shift to higher levels of fracking jobs is therefore not a cause for celebration, although it does benefit a few key states—North Dakota, Texas, Pennsylvania, Colorado, and New Mexico.

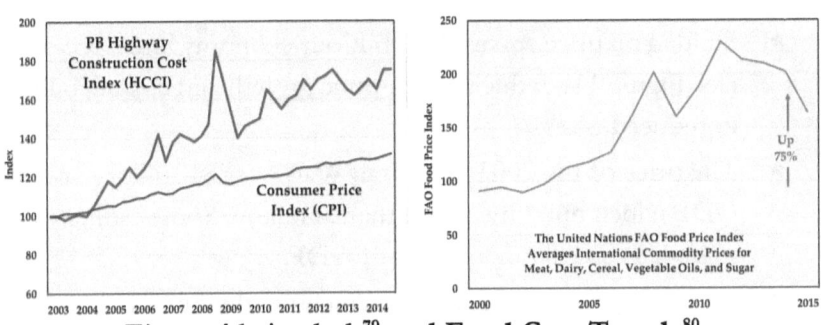

Figure 16. Asphalt[79] **and Food Cost Trends**[80]

Higher oil prices have also increased asphalt road construction costs. Those costs have increased by 40% over inflation in the past decade as shown in Figure 16, and the federal Highway Trust Fund has been annually running dry since 2008[81]. Fifteen percent of roads in the U.S. are in poor condition, and over 10% of bridges are structurally deficient[82].

Oil prices also affect food prices, as oil based transportation is used for harvesting, and for shipping food to market. Global food prices have increased by over 75% since 2002-2004[83].

The U.S. government, which estimates U.S. food price increases since 2000 of about 30%, wraps these food prices into the Consumer Price Index, a common measure of inflation. This makes these increases less visible, as we take it for granted that prices will increase over time.

The direct cost of oil price increases, road maintenance, food cost increases, and increased raw material cost of plastics do not include several hidden costs of American dependency on oil. These are well-documented in Amory Lovin's book "Reinventing Fire"[84] and elsewhere. Hidden costs include:

- $329 billion[85] in 2012, or 1.1% GDP, spent on imported oil that went abroad. "Some portion of this pays for state-sponsored violence, weapons of mass destruction, and terrorism." [180]
- Sudden oil price spikes that shift our economy into a recession (see Figure 11), reducing economic growth and costing billions in reduced taxes.
- The price of the 2003-2010 Iraq War, over $1 trillion, or 3.5% GDP (albeit one-time rather than annual)[86]. Some estimates are as high as $3 trillion, or 10% of GDP.
- The unknown but small percentage of the $625 billion 2013 defense budget spent on protecting oil infrastructure and shipments worldwide.

- Wealth concentration in oil countries "hinders the spread of democracy" and oil countries "tend to have more corruption, repression, and inequity" [180]; many are unstable.
- Environmental cleanup costs (oil spills ($42 billion by BP for Gulf oil spill[87]), etc.)

Overall, the worldwide shift in the last decade from less than $20 a barrel oil to $35 to $100 a barrel costs us about 3%-5% of GDP each year[88]. It is a key reason we have had difficulty shaking off the recession despite our Saudi level oil production income of 1% GDP. These increased costs offset our long term inflation-adjusted 2-3% average growth rate.

So even if you deny that global warming and CO_2 are connected, we need to decarbonize our economy to counteract recent fossil fuel cost increases and allow our economy to grow -- as long as the alternatives are less expensive than fossil fuels.

The System is Rigged — But It's Not Your Fault

What happens if we continue to put our head in the sand relative to CO_2 pollution, and continue to pollute for "free"? Higher sea levels (eventually), a sixth mass extinction event, including ocean food chains and saltwater fish[89], potentially more damaging and/or weirder storms (drought, fire, etc.)[90], water shortages and deluges, and millions of people displaced on the coasts. Sea levels will rise slowly over centuries, given how much ice must melt to change sea level. The slow pace may keep refugee levels manageable or at least similar in size to previous mass migrations. However, there may be pockets of refugee concentrations that will be larger—i.e. in Bangladesh, Florida, East Coast cities, etc. And this does not factor in war, which may increase in the face of climate change[91] (for instance, a severe drought may have exacerbated the Syrian war[92]).

We also feel helpless, because we know that China has overtaken our total emissions output, India is not far behind, and we can't reasonably tell the developing world to stop using the fossil fuels that drove our own economic growth as they struggle to escape poverty.

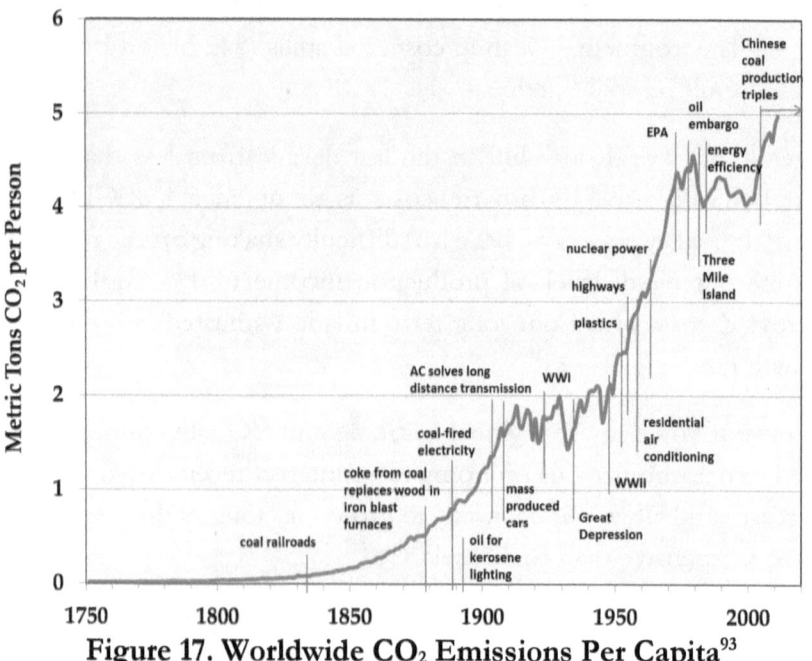

Figure 17. Worldwide CO$_2$ Emissions Per Capita[93]

Figure 17 shows worldwide emissions per capita of 5 metric tons of CO$_2$ per person, and it is not fair or just to tell developing economies to cut their emissions when we emit 17 metric tons of CO$_2$ per person. Rapidly growing developing-world emissions means that anything we do unilaterally will have zero impact on worldwide emissions overall. We could spend a lot and get nothing to show for it, while fueling others' free-riding economies.

If the U.S. pays money to reduce our emissions —whether by taxes, cap and trade, renewable subsidies, R&D funding, forgoing consumption, improved efficiency, or other methods—the costs are immediate, while the benefits are long term and uncertain. While I like to think that I will sacrifice for my grandchildren, in reality people often resist making these kinds of sacrifices[94]. We don't even sacrifice for our own future, as too-low 401K contributions show[95], never mind for our grandchildren. If the costs are high and the benefits in doubt, we will procrastinate until we can no longer ignore the problem when it will be too late.

For all these reasons, the 2015 U.N. climate summit was a failure, similar to all the ones that have occurred over the last 23 years. While it succeeded in obtaining nominal agreement, it set up artificial, nonbinding and unenforceable variable targets, so each country will do what it would otherwise do under a "business as usual" plan, rather than what needs to be done. For example, the promised 27% U.S. emissions cut has its starting point set in 2005, when our emissions were at a maximum. So this promise is actually a pledge to cut emissions only 10% in 10 years. We are already halfway there because of coal to natural gas switching, and this 10% in 10 years promise simply follows the U.S. economy's long term decarbonization trends. We will continue to do nothing.

And it's not our fault—it's the system, and the size of the problem.

Lean Engineering: Making Invisible Waste Visible

To solve the problem, the first step under Lean Engineering is to make the invisible waste visible. Then we can measure the waste, and look for ways to eliminate it.

One of the biggest problems with global warming in general is that both it, and the CO_2 that produces it, are invisible. And we tend to ignore anything we can't visualize.

For example, if we consider the entire economy of the United States, and add up all of the planes, cars, appliances, houses, Christmas presents, books, clothes, roads, and stuff we purchase each year, and weigh it all, which item that we produce, import, or export weighs the most?

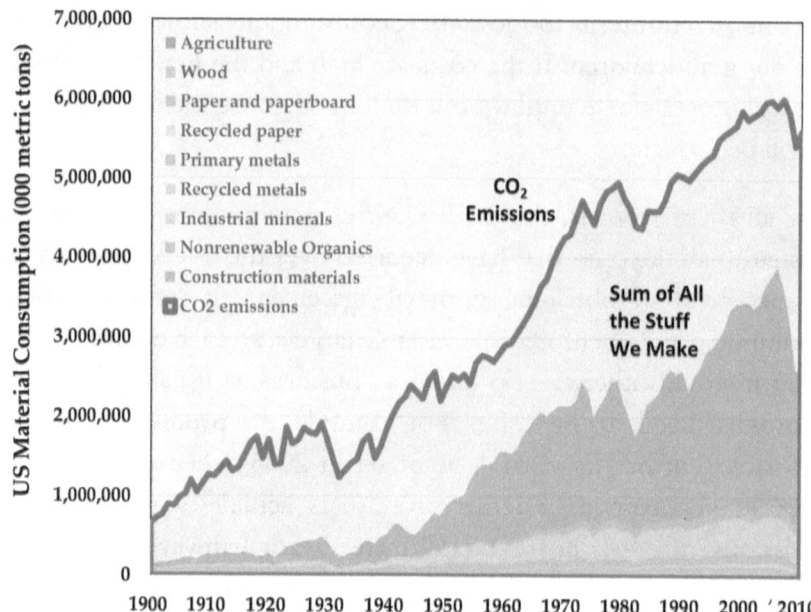

Figure 18. U.S. Annual Top 10 Commodity Consumption[96]

As you can see from Figure 18, the answer is carbon dioxide. If we exclude construction (houses, roads, buildings, etc.), carbon dioxide production exceeds durable goods (cars, refrigerators, etc.) by 5 times[x].

So, really, Americans are very, very good at producing carbon dioxide. All of our worldly goods are the tip of this invisible iceberg.

This graph explains why, even if CO_2 turns out to have absolutely nothing to do with climate, that we do have a CO_2 pollution problem—a rather big one. Imagine if an evil Marvel mastermind was dumping 5,200,000,000 metric tons of glop into our lakes and oceans— wouldn't there be an outcry? Fines? Jail? War? For comparison, the U.S. government estimated the recent Deepwater Horizon oil spill comprised 690,000 metric tons of oil[97], and cost BP

---

[x] Much of our consumer goods manufacturing is in China and other countries, and this graph does not include our portion of foreign emissions for imported goods manufacturing or imported goods material consumption.

over $25 billion. While, yes, it is a very big atmosphere, it is also a very big assumption to think that Mother Nature will clean up this mess for free.

Morally, we need take action to clean up this mess, whether you are following the dictates of the Pope's "Laudato Si" on "Care for Our Common Home," the Jewish "Rabbinic Letter on the Climate Crisis"[98], the Islamic Declaration on Global Climate Change[99], or conservative Christian arguments that doing nothing is not good stewardship of the Earth that God granted us dominion over[100].

But even non-believers should have learned in kindergarten to clean up their messes[101]—even messes left by parents and grandparents. In this aspect, CO_2 does not differ from any other type of pollution.

While we do have significant natural gas reserves that we can burn as a substitute for coal and oil (to some extent), we cannot afford to do nothing about CO_2 pollution. As fossil fuel use increases, the inexpensive sources are running out[102]. Increased fossil fuel extraction costs that reduce EROEI[xi] cost five times as much as immediate climate change adaption and mitigation. The longer we wait, the higher these costs will go. Maximizing our prosperity therefore requires shifting toward less expensive CO_2 pollution strategies as articulated by the "Economic Law of Pollution."

The rest of this book focuses on how lean engineering can make American climate action affordable, what America needs to do to speed up development of technological solutions to reverse the world's CO_2 emissions, and how we can eliminate barriers to action.

[xi] EROEI = "Energy Return On Energy Invested." It equals "energy output" divided by "energy input", and is therefore a measure of the efficiency of energy extraction of an energy source. If we use 1 barrel of oil to dig up 100 barrels of oil to use, this is an EROEI of 100, and this oil can fuel our society. If we use 1 barrel of oil to obtain only 1 barrel of oil to use, this is an EROEI of 1, and the oil is no longer a source of energy. See Appendix F for further discussion.

III

Changes To Activate the Market

"There ain't no such thing as a free lunch."
— Robert A. Heinlein, The Moon is a Harsh Mistress

No Free Lunch

The United States needs to re-examine its current policy of allowing other countries to pollute for us, especially with heavy manufacturing. We rationalize this by saying, "well, we can't dictate what environmental regulations those in the developing world will adopt," and this is true. But this policy costs us a lot more than we realize, and is another example of "invisible waste" that we can first bring to light, then reduce.

What happens when humans pollute for free?

A good example is China, which applied few environmental restrictions as it developed, and to which the U.S. has outsourced much of its manufacturing. Today, despite incredible economic growth and the boot-strapping of 1 billion people out of poverty, China is highly polluted: Beijing is unlivable according to its mayor[103], 40% of arable land is polluted[104], cancer is the leading cause of death, air pollution kills 350,000 to 500,000 people each year[105], and 60% of underground and 30% of surface waters are deemed "unfit for human

44

contact"[106]. Lead levels in Beijing's primary reservoir are 20 times the World Health Organization (WHO) standard[107].

The cost of cleaning up this pollution will be enormous, as the 4th facet of the Economic Law of Pollution applies — China can only use the most expensive option, cleaning up afterward. Air pollution cleanup estimates[108] by Chinese officials vary between $176 billion and $817 billion; water treatment bills vary between $112 billion and $650 billion[109].

However, these official Chinese estimates are likely very low for political reasons. Experts estimate it will take at least 500 billion dollars per year for decades to adequately clean up the environment[110]. This is equivalent to 5% of the 2014 China GDP of $9.5 trillion[111]. In the U.S., the benefits of environmental regulations exceed costs by a ratio of more than 4 to 1[112], and costs are low, around $21 billion[113], or less than .1% of GDP.

So, if the cost of U.S. environmental regulations, following the Economic Law of Pollution, is over 20 times lower than the oblivious haphazard pollution policy followed in China, who will pay for the higher cost of China's pollution? China? Or U.S. and European consumers?

U.S. and European consumers. As a major trading partner, China sends 20% of its exports to the U.S.[114]; and EU trade patterns are similar[115]. This will cost .2% of U.S. GDP in the form of higher prices on goods for decades—a $40 billion[116] dollar annual drag on our economy.

It is stupid to think that we can freely ship our pollution elsewhere, assuming that someone else will clean it up. And this does not consider the effects such pollution has directly on us, such as Chinese coal plumes that increase Los Angeles smog[117].

Given these economic realities, let us recognize what a powerful *economic* force the EPA is, at least regarding reducing pollution cleanup costs. Its Pareto[xii]-focused methodology of cleaning up pollution through prevention is effective and inexpensive. It is bureaucratic and inefficient, it may go overboard or inadequately consider costs vs. benefits, but nevertheless we need to thank the EPA for some of America's prosperity.

Free trade has been a massive enabler of poverty reduction worldwide, and has created a more efficient global economy as capitalism spreads. But shipping our jobs to foreign shores without EPA protections creates a mess that will cost us 20 times more to clean up later, especially as the world shrinks in this era of global corporate hegemony. It is a false economy, and is not free.

If we had a magic wand, the U.S. should impose EPA regulations on all pollution in all countries. Doing so would save us and the world trillions of dollars in future cleanup costs, and reduce compliance costs and uncertainty for global corporations.

While we don't have the power to dictate to other countries, we *do* have the power to dictate to ourselves. Therefore the U.S. could impose "EPA-violation" tariffs on imports from countries that don't meet EPA air, water, toxic substance, and waste pollution standards. But rather than using the money generated by this tariff ourselves, we should donate it back to the EPA equivalent in foreign countries for them to spend it on direct pollution controls, prevention strategies, etc., so as not to hamper free trade (which is critical for continued worldwide economic growth).

If such tariffs are phased in gently and predictably, the impact on consumer prices will be low, because the cost of EPA regulations,

---

[xii] The Pareto principle is a rule of thumb: 80% of the problem is due to 20% of the causes. In a lean engineering context, it is a primary guide to "what's next" for continuous improvement.

spread out over all products, is actually tiny (i.e. .3-2% GDP[118] compared to the 18% GDP we spend on health care). The money generated would assist the developing world and China to clean up their own messes before they are made, and promulgating EPA regulations worldwide would more evenly tilt the manufacturing playing field toward the U.S., helping repatriate jobs lost due to environmental regulation avoidance.

Such a unilateral tariff would eliminate the need for international agreements, and may require some modification to WTO agreements. There would be a lot of dickering about the details of when and how much, which products, etc., but a "no exceptions" philosophy should ensure that pollution standards are applied uniformly to all products sold in the U.S.

But isn't EPA over-regulation is killing our economy? How can promoting EPA regulations worldwide save the U.S. and the world money?

a) *How could we enforce/monitor other countries environmental spending? (i.e. the giving it back to that countries EPA-equivalent)* We cannot manage other countries governments, so we cannot effectively enforce their spending on prevention or pollution control equipment. In cases of corrupt governments, tariff funds can be given to monitored NGOs to ensure that the money is used to further environmental prevention and cleanup. For places NGOs cannot effectively operate, funds would go to other countries cleanup efforts.

b) *Isn't this just equivalent to raising protectionist barriers, decreasing international trade, inviting retaliatory tariffs, and costing economic growth like Donald Trump's recent proposal to raise tariffs 30-60% on China?* Yes, except that the tariff levels will be very small (i.e. .5-1%), as explained later in this chapter.

c) *If it is this small, why bother?* Let's assume China pays 500 $billion annually to clean itself up (as above) for at least 20 years (given the scale of the mess and typical cleanup times). Let's conservatively further assume that proportional to population, 3 other China-equivalent clean-ups occur over the next century as the developing world copies China's "economy first w/o pollution controls" policy. Discounting China because their expensive "clean up afterwards" is already locked in, these cleanups will total $30+ trillion dollars. Imposing a .5-1% tariff on US (and likely worldwide per b above) goods would cost ~$.4-.8 trillion annually; over 20 years this would cost $12-24 trillion, which is much less expensive than $30+ trillion. The world would get an additional $6-18 trillion in GDP growth.

Eliminating the Other Free Lunch

As with any other pollutant, the Economic Law of Pollution can be applied to CO_2 to eliminate the falsely "free" use of our atmosphere as a dumping ground. In our society, the EPA eliminates pollution externalities by applying a variety of penalties, including fines, for violation of pollution standards. The "polluter pays" principle is fair and has a sound legal basis. This approach will also work to solve global warming, while allowing economic growth to occur, just as it has in the past for other air pollutants such as SO_x (which forms acid rain) and NO_x (a component of smog).

Instituting pollution fines is the cheapest, simplest way to halt our current false economy of CO_2 polluting for free, because there is no free lunch in this world even if we have gotten away with it for a few hundred years. Fines should be introduced slowly and predictably, as the problem didn't arise quickly and won't be solved quickly. With adequate advanced warning, manufacturers and consumers can adjust relatively painlessly (based on history), and the EPA should continue to use 80:20 Pareto analysis to focus where to apply pollution taxes to be most effective. These fines will reward entrepreneurs and corporations for inventing cleaner energy sources, and will allow technology to solve the problem of global warming, which it can't do now.

What level of fines makes sense, on a dollar per metric ton of CO_2 basis? Similar to fines levied for other air pollutants, such as SO_x, NO_x, etc., fines should be set only as high as necessary to enable cleanup. A subsequent chapter will outline specific actions we can take to reduce emissions substantially, and there appears to be at least one method of cleaning up our emissions that will cost $20 to $25 a metric ton. This is on the low end of the large variety of carbon pricing proposals and methodologies shown below. The wide spread in estimated costs reflects different assumptions regarding discount rates[xiii], estimates of future global warming damage, and what costs should or should not be included.

Table 2. Carbon Price Proposals

Source	$/metric ton
EU carbon trading allowance	$7-8
CA carbon allowance futures	$13
AU carbon tax (repealed)	~$24
British Columbia carbon tax	$39
U.S. government policy, 3% discount rate (DICE/GCM models)	$40
Exxon internal (2015)	$80
Stern review 2006	$86
Social cost of carbon Stanford study (2015), includes long-term lower growth rate impacts on developing countries	$220

Twenty-five dollars a metric ton is affordable, as it represents less than 1% of U.S. GDP. In a subsequent chapter, it is shown how we can

---

[xiii] A discount rate is a financial technical term for the interest rate used to discount a stream of future cash flows to their present value. At 0%, money today is worth the same as money tomorrow. At 7%, I value money today significantly higher than money I receive many years from now. In general environmentalists believe very low (~.1%) discount rates are appropriate; economists believe 3-5% makes more sense; and corporations tend to make decisions using higher discount rates (7-10%).

further eliminate invisible waste in our economy to save at least 50% more than this, or 1.5% GDP, to realistically afford to get started.

Fines, Carbon Tax, or Cap & Trade?

How would $25 a metric ton pollution fines differ from the many proposals for a carbon tax, or cap and trade? And how can we avoid the mistakes of the past, including the ones made in Australia, which we may repeat?

Global fossil fuel companies are OK with a carbon tax if such a tax is "revenue neutral"—which reduces its impact on their bottom line[119]. Cap and trade schemes may be more politically palatable in the U.S. because they are not a "tax." But what makes the most sense?

Economist Nordhaus says "those who burn fossil fuels are enjoying an economic subsidy -- they are grazing on the global commons and not paying for what they eat."[120] His primary solutions, offered from an economist's perspective, include (1) establishing a carbon price through a tax on carbon, or (2) equivalently establishing emission limits via a cap and trade scheme.

Carbon Tax Solutions

With a carbon tax, taxes are levied at the point of entry in to the market (i.e. at the oil well, coal mine-mouth, or natural gas well), but the level of emissions is uncontrolled and unknown. To make a carbon tax revenue neutral, William Nordhaus (and others) propose to use direct rebates to consumers, and, from an economic perspective, this is claimed to be the cheapest way to reduce emissions. A big issue is that because the poor spend a much larger proportion of their income on energy costs, they would be disproportionately impacted, and carbon taxes are therefore regressive … so rebates would be critical to ensure that the poor are not harmed.

Another issue regarding carbon taxes is that they need to be applied universally to all countries, otherwise polluters will move operations (to the extent possible) to low carbon tax areas, negating its purpose.

Cap and Trade

Cap and trade has almost the same effect as a carbon tax, but in reverse. The level of emissions is controlled (through permits), and the price of CO_2 emissions permits may float. U.S. politicians and industry favor cap and trade because higher complexity allows for more loopholes and lobbying potential, and in the past, industry got free initial permits, which reduces actual changes to be made by the industry[121].

The EU's Emissions Trading Scheme has shown it can be successful at nominally reducing emissions. However, this version of cap and trade has also struggled with grandfathered windfall profits, inefficient permit allocations, leakage emissions, carbon price crashes and volatility, computer hacking/security breaches, and VAT fraud. Overall, emissions did not reduce much[122], for a fraction of 1% of GDP[123] of the EU states. Offsets and allowances have also diluted its effectiveness. For all these reasons, economist Nordhaus favors a carbon tax[124].

Historically, experiences with both schemes vary worldwide:

(1) The EPA reduced acid-rain sulfur emissions by 80% since the 1970s in the U.S. via a cap and trade system.

(2) The Canadian province of British Columbia's experience with a revenue-neutral $39 a metric ton carbon tax has been positive[125], though their 86% dependence on hydropower electricity generation sharply blunted its impact on electricity prices[126].

(3) Australia, which generates nearly all its electricity using coal, imposed a lower $23 a metric ton carbon tax that it subsequently repealed due to political unpopularity, rising electricity prices, and mining and fossil fuel industry

opposition. Ironically, AU coal prices doubled between 2000 and 2015 (spiking higher and dropping again in between). This is one of the invisible causes of increasing electricity prices by about 30% —but the 5%-10% electricity price increase due to the carbon tax got the blame.

In summary, cap and trade schemes are complex and less effective at reducing emissions due to offsets, allowances, and carbon price volatility; while carbon taxes have a tendency to be repealed and/or suffer from political volatility, and lead to emissions volatility. The history above shows that both schemes can work well if the impacts on society are low. But this also means they are likely to be relatively inadequate (i.e. not reduce emissions that much). So why bother?

The goal of introducing low level pollution fines equivalent to a carbon tax is *not* to reduce emissions. It is to get us to pay attention and care about CO_2 emissions, and activate our market economy to pay entrepreneurs, investors, and corporations appropriately to develop technical solutions. It is these technical solutions that will reduce emissions, not carbon fines themselves (although fines will have a small effect by raising prices slightly). Without a price on carbon, the market will not care or change CO_2 emissions.

Low level fines that increase fossil fuel prices will have a secondary effect that results in small reductions in usage (driving less, merging trips, etc.). For example, this has occurred naturally in the U.S. since 2005 as consumers responded to higher oil and gasoline prices, shown in Figure 19. Gasoline prices went from under $2 a gallon to $4 a gallon, and, more recently, up to $3 a gallon[127].

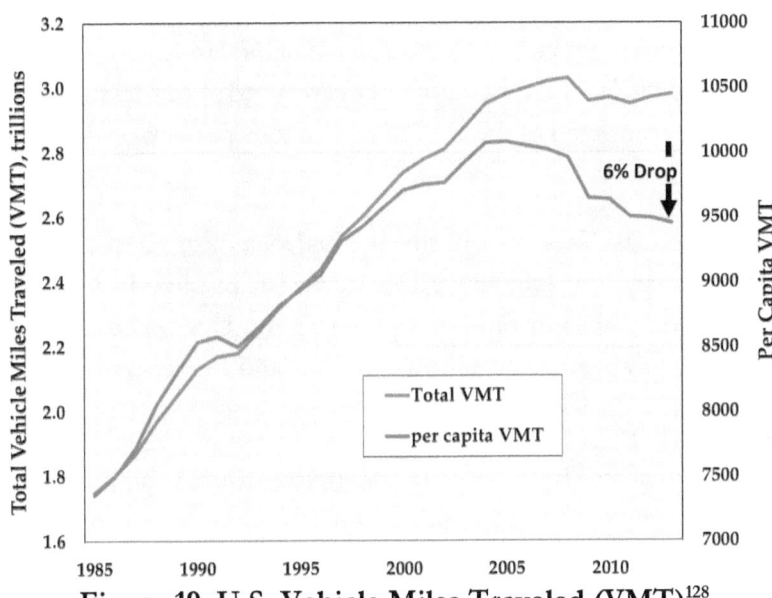

Figure 19. U.S. Vehicle Miles Traveled (VMT)[128]

But, this "limiting usage" method (which also goes by the names "drive less," "shop less," "use less," "eat less," etc.), while effective and inexpensive, also has economic growth implications. Because economic growth and GDP measure consumption, if we limit usage, we reduce our economic growth and wealth. America's implicit orientation toward economic growth and consumption is therefore one reason that carbon taxes or equivalent cap and trade are unpopular. We like to consume, and carbon dieting is just not fun. In addition, because our country is so spread out, there is a limit to how much we can easily reduce. Therefore I excluded the "limiting usage" method from the Economic Law of Pollution, even though it is the least expensive method of reducing emissions because it costs nominally nothing[129].

We could raise prices on carbon a lot by imposing high gasoline and energy prices. In Europe and Japan, higher energy prices have been a mixture of natural pricing and regulation. In Japan, the island imports nearly all forms of energy, so prices are naturally high; in Europe, additional energy taxes are imposed. Higher prices, along with geography and other cultural factors, have helped the EU and Japan to reduce their "tons of CO_2 emitted per capita" to about 10 metric tons per person, compared to the U.S. 17 metric tons per person (see Figure 2).

But because fossil fuel usage forms the bedrock of modern economies per White's law, these higher prices have also resulted in lower long term economic growth rates compared to the U.S. Since 1970, U.S. growth rate has been 2.4%, Germany 1.6%, and France and Japan less than .2%[130].

If low level carbon fines become a stepping stone to high fine levels that are the equivalent of high carbon taxes, we would reduce our economic growth and living standards. This market distortion should be avoided by making sure that the EPA does not charge more than cleanup costs, as is their typical practice (with punitive exceptions for civil and criminal violations).

Similar to carbon taxes, however, market distortions will occur if we allow exceptions to carbon fines in the U.S. In particular, EPA's Clean Power Plan only applies to large power plants. As a result, power plants that burn less than 250 MMBTU/hour heat input of fossil fuel are not regulated[131]. Over time, utilities might favor smaller power plants to take advantage of this exception, and this could distort the market. Similarly, because the Clean Power Plan applies only to power plants which constitute about 40% of U.S. CO_2 emissions, and not to all carbon emissions throughout the U.S. economy, this will favor some economic sectors over others.

In a political climate that has rejected any kind of CO_2 tax, fines, and/or regulations, one has to start somewhere, and it makes sense to focus on the largest power plants first as the EPA has done. However, these loopholes need to be closed so that *all* CO_2 emissions have the same rules to prevent long term market distortions.

Fixing Other Market Distortions

Aside from the large "CO_2 emissions for free" distortion addressed above, other market distortions are plentiful[132] and continue to warp energy markets:

a) There are federal gasoline taxes in the U.S., but not coal taxes[133].

b) The Clean Air Act allowed power plants that existed prior to the act's passage to continue to pollute without emissions controls[134]. This compromise led to the utility industry extending the lifetimes of these grandfathered plants long past their original 30 year design lifetimes[135], with many inefficient highly polluting plants now averaging over 50 years of age[136].

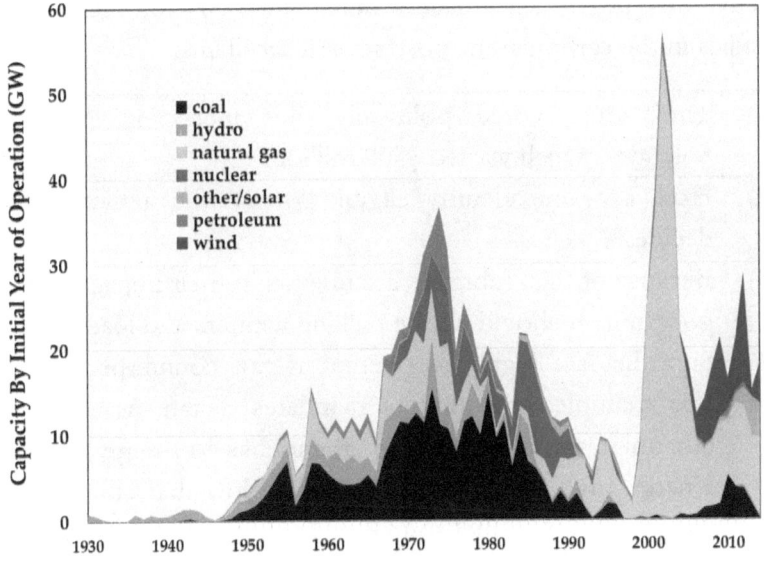

Figure 20. U.S. Electricity Generator Vintage[137]

c) From 1978 to 1987, the Powerplant and Industrial Fuel Use Act prohibited the use of natural gas for electricity generation[138].

While the EPA's Clean Power Plan may indirectly address (a) and (b) above, we need to look at other market distortions to allow unfettered competition to allocate resources, reduce costs, and promote economic growth. Two large invisible distortions still present in markets include subsidies, and energy cost myopia.

Subsidies

Carbon taxes are an attempt to fix a market externality by applying a penalty, charging one directly for the negative effects of pollution (and then giving it back at tax time with revenue-neutral plans). Subsidies, in contrast, spend your (tax) money to encourage good behavior, rather than applying a penalty.

Psychologists will tell you that using positive feedback is much more effective than negative feedback for changing behavior. Nevertheless, subsidies in the carbon arena pose several problems:

1) They are favored politically, but inefficient (remember Solyndra, which wasted $500 million?)

2) How can one identify eligible low-carbon activities? Who decides?

3) Because of this, subsidies are uneven in their impacts, and the government should not be picking winners and losers.

4) Subsidies are highly ineffective, if not counterproductive[139]. For example, ethanol fuel mandates dictate usage of corn ethanol in the U.S., but corn ethanol has an Energy Return on Energy Invested (EROEI) that is so low that ethanol is not worth using as a fuel source. This impoverishes us all while distorting food markets and land use.

5) Subsidies tend not to be sustainable, and/or depend on political winds. For example, the wind industry has suffered from incentives turning on and off periodically for the last few decades[140], leading to market uncertainty that has alternately slowed and spurred market growth.

6) Subsidies distort market supply and demand, leading to warped high or low prices. For example, China subsidized solar production facilities in 2007, giving low (or no) interest loans for capital equipment for manufacturing. PV module[xiv] supply exploded, cratering PV module prices and driving many U.S. and European PV module manufacturers out of business. Over half of worldwide solar production shifted to China[141] in the last decade, leading to U.S. anti-dumping tariffs[142]. Chinese companies, generally spending minimally on research, subsequently purchased many failed companies to acquire for pennies on the dollar new technology that was developed in the U.S.

7) Subsidies favor incumbent technologies. Fossil fuel subsidies in the U.S. have been dramatically higher than for renewables over the last 50 years[143], and fossil fuel taxes are uneven among sources. In the U.S., over $20 billion[144] in fossil fuel subsidies include:[145]

    i. Federal exploration subsidies (expensing of drilling costs, two-year amortization period for geological and geophysical expenditures, 70% deduction of domestic exploration and development costs for hard mineral fuels, etc.)

    ii. Federal production subsidies (percentage depletion adjustments to property tax basis)

    iii. Consumption subsidies (low income home energy assistance programs)

    iv. Cleanup costs (BP claimed tax deductions for the Deepwater Horizon spill in the Gulf of Mexico, for example)

    v. Similar individual state government subsidies

    vi. U.S. financing of overseas fossil fuel projects[146]

---

xiv A PV module is a group of connected solar cells that produce electricity. Typically, a PV panel or module is about 2 feet by 4 feet in size, one-inch-thick, weighing less than 50 pounds. Many modules are mounted in arrays on the ground or on a rooftop.

vii. U.S. federal R&D subsidies (see Figure 22 for example), which helped fracking technology commercialization and the U.S. resurgence as an oil and gas producer.

8) Subsidies can become more perverse over time. For example, the 1998 Direct Subsidy in the Farm Bill, repealed in 2012, paid large sums of money to wealthy farmers over decades, rather than helping smaller farmers as intended[147].

Renewable and Fossil Fuel Subsidies in the U.S.

Digging deeper, a number of studies track renewable and fossil fuel subsidies[148], which is a murky and controversial task. As Figure 21 and Figure 22 show, fossil fuel subsidies have been much larger than renewable subsidies. However, this picture can easily be distorted if one looks at short periods of time—say, emphasizing recent ARRA funding where renewables subsidies briefly exceeded fossil fuel subsidies—or if one looks at subsidies per energy unit; low usage of renewables by the market makes "subsidies per energy unit" much higher for renewables compared to fossil fuel[149]. As but one illustration of the controversy in this arena, note the differences between Figure 21 and Figure 22.

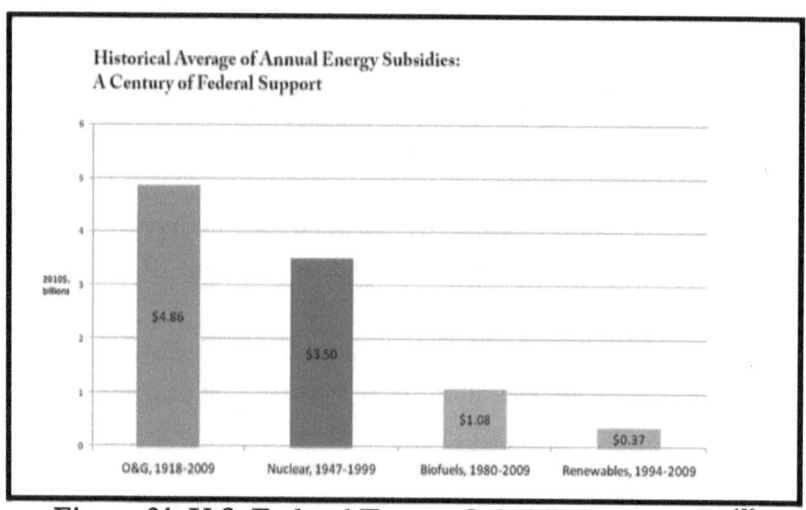

Figure 21. U.S. Federal Energy Subsidies, one study[150]

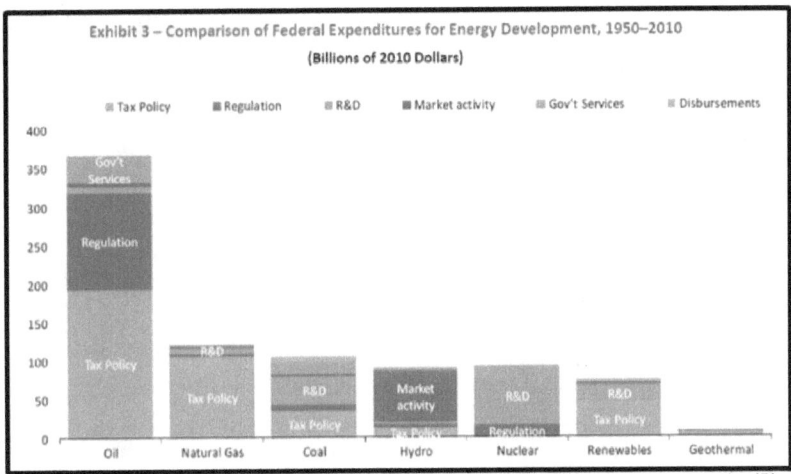

Figure 22. U.S. Federal Energy Subsidies, second study[151]

The first study attempts to estimate the impact of the Price-Anderson Nuclear Industry Indemnities Act of 1957, which limits nuclear industry liability when accidents occur; the other, sponsored by the nuclear industry, omits this subsidy. Neither study measures the costs or subsidy implicit in ignoring the costs associated with nuclear waste disposal. While subsidies may be good psychology because they work, and good politics because they are a source of influence, when subsidies are introduced, the market reaches conclusions and outcomes that are flat-out wrong. Cases in point include ethanol, and nuclear power.

Ethanol Subsidies

Corn based ethanol was first blended into gasoline in response to gasoline price spikes in the 1970s. Corn ethanol subsidies of 40-50 cents a gallon were enacted in 1978[152], renewed in 2004[153], and eliminated in 2011; however, the federal Renewable Fuel Standard (RFS) still mandates a minimum percentage of corn-based ethanol annually[154]. The planting of about 40% of U.S. corn crops for ethanol on an area roughly the size of California (90 million acres) increases food prices for the world[155], increases our emissions, uses up non-renewable Ogallala Aquifer water, and is more costly and less efficient than gasoline. Ethanol regulations cost about .5% GDP, and should be repealed[156] to allow more conversion of crop land to forest (which would decrease emissions). Research into cellulosic and/or algae ethanol should continue, as these might address the core "energy density per acre of land" productivity issue that sharply reduces corn's effectiveness as a biofuel crop.

Nuclear Subsidies

The Price-Anderson Nuclear Industry Indemnities Act of 1957 states that U.S. taxpayers will pay liability costs if a nuclear accident or incident occurs, thereby subsidizing insurance costs for the nuclear industry. But if the nuclear industry is safe as purported, such liability coverage should not be necessary. The risks of nuclear power plant operation, with more than 50 years of industry operating history, are well known and low.

The problem is that accidents do happen because of human error, as Three Mile Island, Chernobyl, and Fukushima were all preventable[157]. Even in technologically sophisticated Japan, engineers did not expect a 2011 tsunami cutting power to the Fukushima Daiichi backup generation systems, leading to a large leak that four years later was not contained, with radiation reaching the U.S. West Coast[158]. In 1986, Chernobyl killed and sickened many, leading to a large area radiation release in Europe. And in the U.S., while Three Mile Island in 1979 was the most serious, over 50 nuclear reactor accidents have occurred since 1952[159]. Because low risk is not zero, nuclear technology has proven to be not worth the risk because of the massive costs incurred when anything does go wrong. Germany shut down its nuclear reactors after the Fukushima accident, and Japan struggles to get its nuclear plants back into service. In the U.S., higher costs associated with improved safety standards and custom nuclear plant designs, not in my backyard (NIMBY)-ism, and high or unknown decommissioning and storage costs have also rendered nuclear power expensive compared to other power generation options. Only four new nuclear reactors[160] are slated to come online in the U.S. by 2020. They are already 20% over original construction budgets.

Given this safety and cost track record, and the huge cost of incidents ($105 billion for Fukushima[161]), it makes little sense for the public to shoulder liability insurance for this mature industry.

Energy Cost Myopia

Besides subsidies, another market distortion exists in energy markets that economist Nordhaus names "Energy Cost Myopia"[162]. This is a syndrome in which people invest too little in energy efficiency because they under-weigh (or over-discount) future fuel savings. We want our energy savings immediately. One example Nordhaus offers is our choice of whether to purchase a diesel or gasoline car. The purchase of a diesel engine costs $2,000 more, but saves 11 mpg annually. That's $640 per year for the average American, and a 3.1 year payback. But most Americans will purchase a gasoline car, with gasoline cars outselling diesel cars by over 2:1. The reasons vary from poor information to being cash starved to not caring enough.

Not caring enough is exacerbated by low energy prices. If energy prices are low, then total costs for consumers are small relative to our entire annual budget, which leads to apathy and ignoring of energy costs, energy efficiency, and our consumption. The benefits of saving on energy costs are lower than the transaction costs and inertia of doing something different, and consumers invest in upgrades of energy-related equipment for comfort, health, aesthetics, safety, reliability, convenience and status reasons rather than paying attention to energy costs[163].

Lean Engineering Market Activation Summary

Fossil fuels cost increases are four times[164] more than estimated global warming costs, and these costs continue to weigh on our economic growth and will increase over time. Therefore, it makes sense to activate our market economy now to simultaneously solve the problem of finding fossil fuel alternatives while reducing global warming emissions. At present, with a zero price on carbon, technology cannot solve the problem. Market distortions in the energy sector are prompting the market to make incorrect choices—nuclear power is too risky (as it would not exist without federal government liability insurance subsidies) and ethanol policy is unnecessarily increasing food prices and slowing economic growth. To make accurate choices, we need to create a level playing field and let the market decide, removing all barriers and subsidies and allow the best technology(s) to win. The capitalist market has shown itself to be the most effective and efficient way to make technology and cost choices that promote our well-being —if the market accounts properly for externalities.

Unfettered competition therefore includes:

a. Eliminate allowing other countries to pollute their land for the sake of cheaper goods here. The Economic Law of Pollution shows that U.S. consumers will pay 20 times more than necessary for pollution cleanup costs in higher prices on goods for decades, as China slowly cleans up after itself. If we continue on our present course, this same dynamic will play out and pollute Indonesia, India, Africa, and other low wage countries. And then we will pay 20 times more than necessary in these locales, too.

b. While we cannot control our neighbors, we can control what we ourselves will accept. Gently and predictably, the U.S. should impose "equivalent EPA pollution" taxes on all imports that are manufactured without following EPA standards, and give this money back to the EPA-equivalent in the relevant foreign country (and/or monitored NGOs in the case of corrupt governments) to invest in direct pollution controls, more effective pollution strategies and laws, etc. This will also help to repatriate jobs lost due to environmental regulation avoidance.

c.   Eliminate free CO_2 polluting of our atmosphere by allowing the EPA to impose CO_2 fines, as recently established by the Supreme Court. Oil and coal are already too expensive, and the sooner we shift away toward something cheaper, the better. At the same time, this will lead to decreased CO_2 emissions.

    i.   While slightly higher prices caused by fines will stimulate innovation (as occurred with fracking), too high prices will slow economic growth. CO_2 fines should not be higher than cleanup costs, as for all other pollutants. If low level carbon fines become a stepping stone to high carbon taxes, we will lower economic growth, as shown by the EU and Japan[165].

    ii.   For an even playing field, CO_2 fines need to be imposed uniformly throughout the economy on all sources of CO_2 pollution. The current hodgepodge of higher CAFE standards combined with the EPA's Clean Power Plan is a start, but will produce undesirable distortions as some economic sectors are favored over others.

    iii.   Carbon taxes, subject to political winds as shown by Australia, will likely not work in the U.S. Equivalent cap and trade schemes are overly complex and less effective (as the EU system has shown).

    iv.   Slightly higher fines will partially address "energy cost myopia" by causing us to pay attention to emissions.

    v. CO_2 pollution fines should also be part of "b" above, to avoid other countries "free riding" on our efforts, and to allow the U.S. to take effective unilateral action without regard to international agreements.

d. Eliminate all energy-related subsidies, to further level the playing field and allow the market to make accurate decisions.
   1. Repeal corn ethanol RFS mandates, saving about .5% GDP on food, gasoline, and restaurant prices[166].
   2. Phase out fossil fuel subsidies of exploration, depletion, and cleanup cost tax breaks, and U.S. financing of overseas fossil fuel projects (but continue research subsidies).
   3. Repeal Clean Air Act grandfathering of existing coal plants without emission controls.
   4. Repeal the Price-Anderson Nuclear Industry Indemnities Act of 1957.
   5. Eliminate subsidies for the renewables industries.

Let the market decide.

IV

Expensive Fossil Fuels Reduction Plan

"Make no small plans -- for they have no power to stir the soul."
– Niccolo Machiavelli

In the previous chapter, three overarching invisible wastes are outlined of exporting our manufacturing pollution through outsourcing, polluting CO_2 for "free", and energy subsidies distorting market decisions. If we eliminate these wastes, the market will be enabled to find alternatives to expensive fossil fuels while solving global warming.

But what can we do right now to lower emissions growth, if not lower absolute emissions, as scientists say is necessary?

In this chapter, specific actions are examined that a market-enabled economy could take, following a lean engineering approach to reduce expensive American fossil fuels use and CO_2 emissions significantly.

First, let us look at the underlying drivers of our CO_2 emissions to find cost effective solutions that affect these drivers. Second, let us examine what has worked in the past to reduce emissions cost effectively, and how lean engineering principles can direct future research. Finally, the historically effective Economic Law of Pollution is applied to our current CO_2 emissions sources, focusing on the largest emissions sources first.

The result is a short term action plan that can reduce our fossil fuel usage by 35%, and a longer term research plan to overcome the remaining technical and economic barriers that prevent further reduction.

Fossil Fuel Usage and Emissions Drivers

The "I = PAT" equation, from sustainability circles, posits that our CO_2 emissions are the product of three factors: population, affluence/wealth, and technology.

Impact (CO_2 emissions) =

Population x Affluence x Technology

Current models of economic growth attribute wealth and affluence to factors such as capital, total population, and technology (productivity growth). If we substitute these wealth and affluence factors for the Affluence variable above, two factors emerge that are the primary drivers of our fossil fuel emissions:

CO_2 emissions   = Population
  x (Capital, Population, Technology (productivity growth))
  x Technology
Or:
CO_2 emissions   = Population^2 x Capital x (Productivity Growth)^2

Per the 80:20 Pareto rule, let us now focus on population and technology (i.e productivity growth) in the next two sections, because they affect CO_2 emissions the most.

Population Growth

If we look at current and future worldwide population estimates, plotted against worldwide emissions, there is a strong correlation between emissions and population.

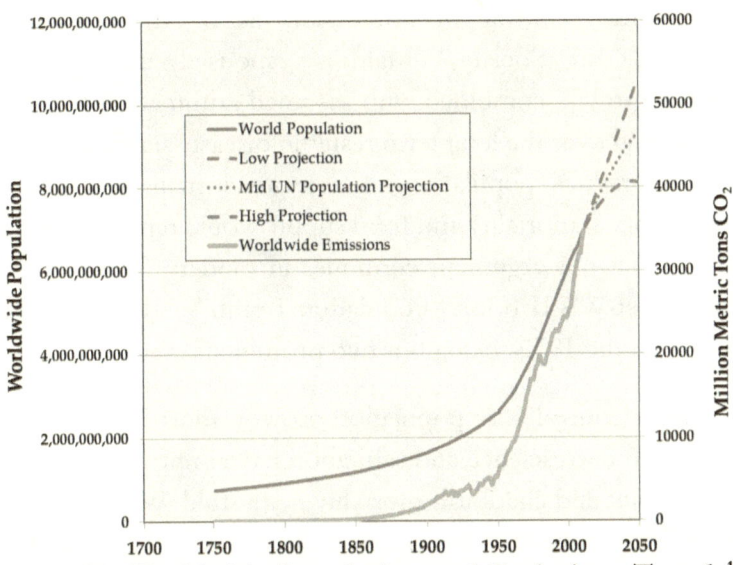

Figure 23. Worldwide Population and Emissions Trends[167]

As an engineer, at the start of researching this section, I had vague thoughts that the population side of the equation —important as a double driver of emissions —would be amenable to the 80:20 Pareto approach taken in the rest of this book. But then, diving in to the over 100 year-history of humankind's attempts to control its population, I found a cautionary tale, as our efforts to do so boil down to only one thing that both works to reduce population and is not morally repugnant.

Along this journey through mankind's attempts to control population, there are numerous perverse subsidy and manipulation pitfalls, with driving forces of racism, imperialism, and eugenics that still haunt us today, even as China has reconsidered its one child law. Below, I summarize this journey, drawing on ideas discussed in Matthew

Connelly's book "Fatal Misconception: The Struggle to Control World Population"[168]. If you are interested in more depth regarding this surprising and fascinating history, his book is a good starting point[169].

The history of population control starts with Malthus in 1798. In his "Essay on the Principle of Population, as It Affects the Future Improvement of Society," Malthus espoused that exponential population growth, combined with assumed arithmetical food supply growth, would over the long term result in disease, starvation, and war that would reduce population to a more sustainable level. Later improvements in mortality and food supply would repeatedly disprove this thesis, but this argument continues in modern forms (with Paul Ehrlich's post-WWII book "Population Bomb" and "The Limits to Growth"[xv] in the 1970s being but two prominent examples).

Three things control U.S. population growth: mortality, fertility, and migration. Modern science and sanitation have improved mortality by reducing infant and childbirth mortality many-fold. Wars, while killing many millions in the last two centuries, have not lowered population, but rather have surprisingly *added* to humanity's population growth, due to the baby booms that occur when wars end[170]. Medical science's knowledge of disease has "over-doubled" human lifespans, both infant and adult, increasing population and reducing mortality. Modern technology has also discovered more ways for us to control fertility through various forms of birth control. While U.S. immigration policy has its problems, it acts to balance the United States' sub-replacement fertility rate, stabilizing U.S. population growth by admitting[xvi] a million[171] legal immigrants per year.

Popular in the early 1900s, eugenics policies try to improve human genetics through genetic screening, birth control, promoting differential birth rates, marriage restrictions, segregation, compulsory

---

[xv] See endnote 429 for complete reference.
[xvi]Note, when recessions hit, migration direction can reverse, as occurred recently in 2008.

sterilization, forced abortions or pregnancies, and genocide. Pioneered in the United States in the 1920s, carried to an extreme by the Nazis and in India, it has since been outlawed, but examples still crop up in the modern world. China's one-child policy is a prime example[172] along with Israel's recently forcing Depo Provera birth control on unsuspecting Ethiopian Jewish immigrants[173]. In the U.S., the government conducted eugenic sterilizations in several states throughout the 1950s.

India launched the first national policy to limit population growth in the 1950s, funded and designed by international organizations. It was an abject failure, as at its root, it comprised telling "the dumb millions ... not what they want, but what is good for them."[174] This American hubris— American because the U.S. funded the population control movement through USAID and the Ford and Rockefeller Foundations —continued through the 1960s, when millions of Indian women were coerced or bribed into accepting IUDs without regard to the risks of these devices.

As shown in the National Security Study Memorandum 200 of 1974, official U.S. national security policy supported population reduction initiatives worldwide, focusing on India, Bangladesh, Pakistan, Indonesia, Thailand, the Philippines, Turkey, Nigeria, Egypt, Ethiopia, Mexico, Columbia, and Brazil. While conservatives in the '80s stopped funding abortion as part of these initiatives, since then the U.S. has continued to spend millions on population control.

This aid has been largely rejected, as USAID now supplies 50 million couples worldwide with family planning[175], which is less than 2% of those living on less than $10 a day. This is the key lesson of the population control movement and its history: Telling other people what to do, or coercing them ("good for them" or not) is both a waste of time and money, and runs counter to America's values of freedom and choice.

The One Thing That Works

Looking closely at the history and the data, Connelly shows that the so-called "demographic transition," where fertility rates fall precipitously when society shifts from rural toward cities (as referenced by economist Nordhaus previously), does not actually exist. Throughout the world and through the ages, fertility falls in *both* rural and urban areas in a region in response to better education of women and reproductive choice, irrespective of income level. Shifting from a rural to an urban economy improves overall wealth as a country develops, but it is the better education and reproductive freedom of women, not the shift into cities, that reduces population[176] —and this is the one thing that works. Figure 24 and Figure 25 show a snapshot of these trends for each country in the world in 2010.

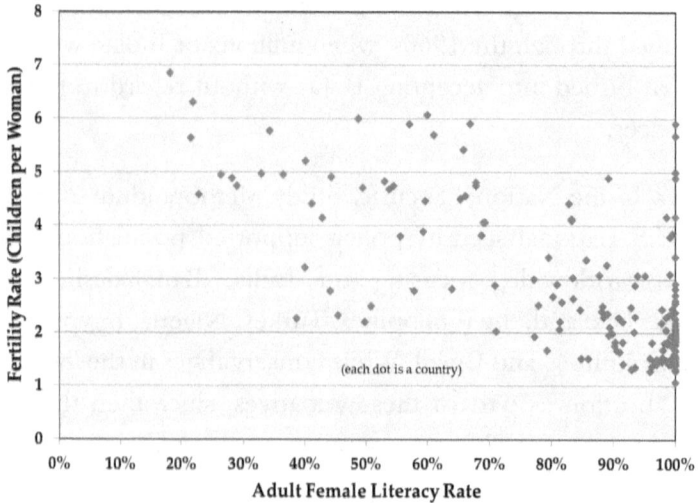

Figure 24. 2010 Female Literacy vs. Fertility Rate[177]

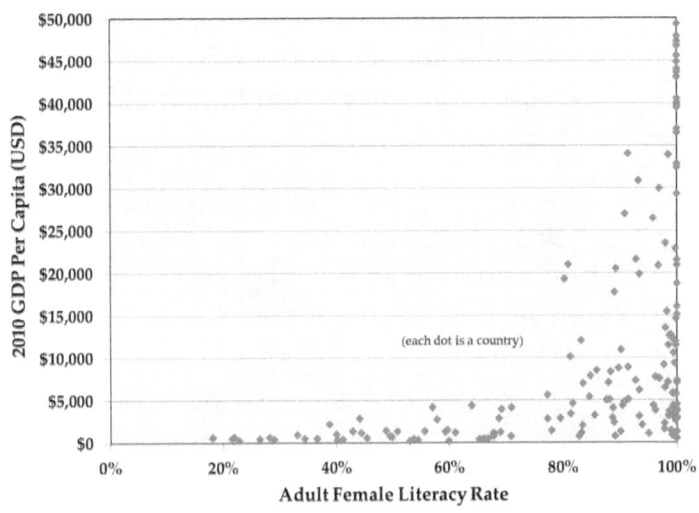

Figure 25. 2010 Female Literacy vs. 2010 GDP per Capita[178]

In developed nations, where the cost of educating and having children is high, birth rates per woman have fallen to less than the 2.1 replacement rate, and are as low as 1.3 in Japan, and 1.6 in Europe.

Table 3. 2013 Worldwide Fertility[179]

Region	Population (2010)	1970 Total Fertility Rate	2010-2013 Total Fertility Rate
	(Billions)	(Average Number of Children per Woman)	
World	6.9	4.4	2.5
Developed Countries (> $10,000 GDP/Capita)	1.6	2.2	1.8
More Developed ($1,000-$10,000 GDP/Capita)	4.5	5.1	2.4
Least Developed (<$1,000 GDP/Capita)	.8	6.8	3.7

The least developed and more developed countries continue to have higher than replacement fertility rates of 2.1, but the spread of capitalism and economic prosperity have literally halved worldwide fertility rates over the last forty years.

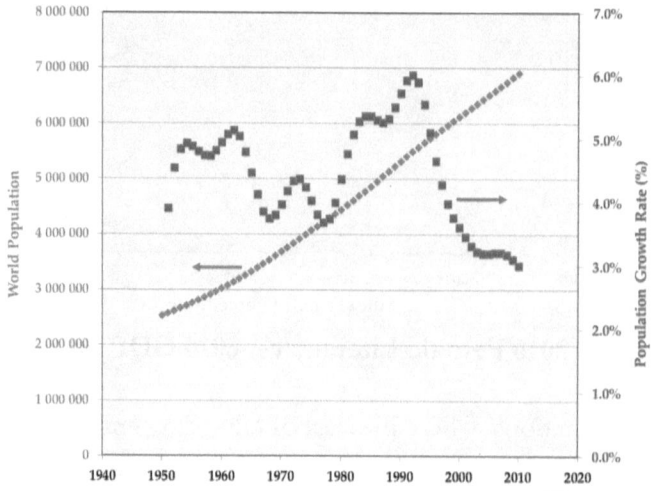

Figure 26. Population Growth Rate[180]

This effect of this can be seen more clearly in Figure 26, where worldwide population growth is shown plotted against population growth rate. Worldwide growth rates topped out in 1990 and continue to fall.

Future Population Growth

Because fertility falls in response to higher education levels for women and greater reproductive choices, the United Nations population division projects that worldwide population will top out at 8 billion (low scenario), 11 billion (medium), and 17 billion (high) respectively. See Figure 27.

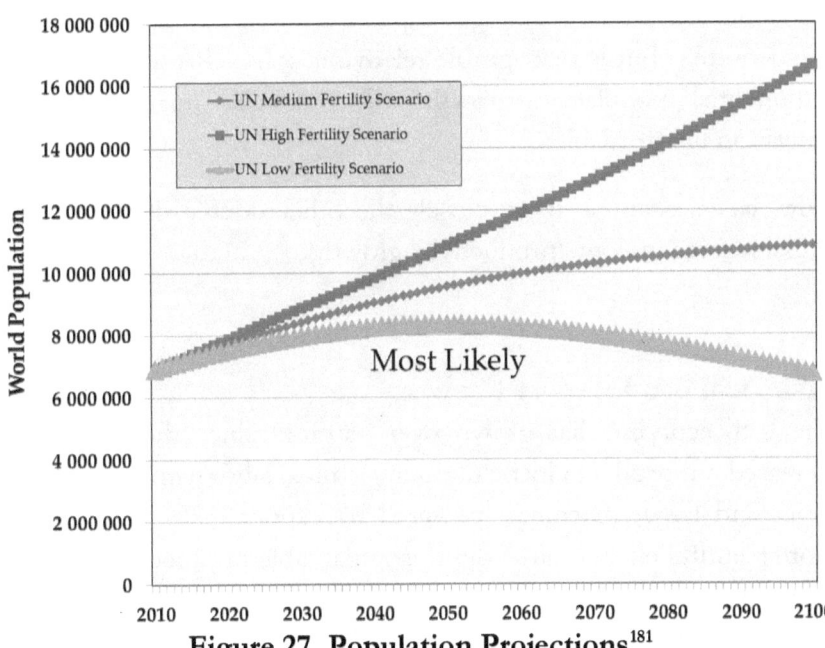

Figure 27. Population Projections[181]

However, digging into the assumptions behind these scenarios, the medium fertility scenario is not credible, because it posits that European, Japanese, and other developed economies will raise their fertility rates from current lows (i.e. 1.3-1.5) to the replacement rate (2.1) by 2050. There is no evidence to support this assumption — rather the reverse. Programs and government efforts to raise fertility rates in these countries have been inadequate and ineffective.

It is therefore much more plausible that the world is actually on the low trajectory or lower. A "Limits of Growth" author, whose standard scenario in the early 1970s accurately predicted population trends for 40 years, has created another population projection out to 2052, which is slightly lower than the UN low scenario. Population is projected to top out at 8 billion by 2040-2050, then decrease.

From a CO_2 emissions perspective, the 14% increase in worldwide population before growth levels off increases the difficulty of cutting

CO_2 emissions somewhat – per capita emissions should decrease to compensate – but is manageable relative to past EPA achievements. Exponential population growth will therefore not make CO_2 mitigation unaffordable.

Now, let us examine more closely the other double driver of CO_2 emissions: technology (productivity growth).

Historical Productivity Growth

The U.S. economy has grown over 450-fold since the 1900s. We increased our productivity by replacing manual labor with horses, then larger and larger machines, powered by exponentially cheaper and more plentiful energy. How did this remarkable productivity increase occur? How can we replicate it to continue to grow our economy rather than having it stagnate because of fossil fuel cost increases?

When we look at the history of doing more with less, this remarkable story of continuous improvement embodies lean engineering. Three key innovation drivers emerge: **efficiency**, which improves technical performance; **scale**, which improves cost; and **paradigm shifts**, which can improve both.

Efficiency

Lean engineering in manufacturing teaches one to look at losses and waste relative to a theoretically ideal, to measure these, and then to continuously reduce them. This method is more broadly applicable to all technical performance realms and is fundamentally how technology performance gains occur. It is a key source of our wealth.

For example, the first steam engine, the Newcomen, had an efficiency of .5%, losing over 99.5% of the energy used to run it. James Watt improved this to 2.5% by using a separate condenser to reduce thermal losses and increase steam pressure slightly to get closer to the ideal. Modern multi-stage steam engines, at much higher pressures, now achieve 10% efficiency. Steam turbines with condensers and steam reheat cycles can achieve 30% efficiency when generating electricity from burning fossil fuels or nuclear fission. This is close to the theoretical maximum, which is limited by a heat engine's temperature, per Carnot's theorem of the second law of thermodynamics.

Similarly, increased automotive efficiency is a story of reducing losses. The development of pneumatic tires reduced rolling friction, regenerative braking saves and reuses the energy normally lost as heat, aerodynamic improvements reduce air drag losses. Internal combustion engine (ICE) efficiency is relatively low, 20%-30%, due to heat engine thermodynamic Carnot limits. Electric engines and drive trains have much higher efficiency than ICEs, leading to 2 times better overall fuel-to-wheel efficiency. Unfortunately, fully electric vehicles still have limited range, safety issues, slow charging speed, limited battery life, auxiliary heating/cooling and high cost/scale issues.

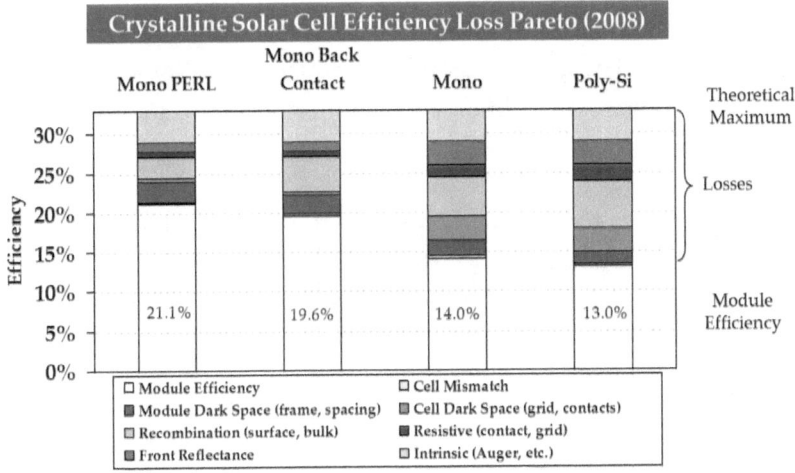

Figure 28. Solar Cell Efficiency and Losses[182] (Illustrative)

A final example is in solar electricity generation. One of the most efficient commercial solar cell designs is made by SunPower Corporation in California. The company achieved higher efficiencies by examining the losses of a more standard poly-silicon (Poly-Si) or monocrystalline (mono) solar cell, and redesigning the product to reduce these losses, relative to a theoretical maximum as shown in the Figure. The process they used is:

1) Model the losses, to enable full understanding
2) Measure the losses, to improve modeling accuracy
3) Improve the design, by reducing losses

At all times, the cost vs. efficiency of any change is tested and modeled to ensure that any design change is warranted.

Scale

Improvements in scale reduce costs, and there are many ways to improve scale. Larger or wider equipment, reduced size of active components or concentration, higher production volumes, and time savings innovations are all examples of improved scale. To improve scale, lean engineering teaches us to measure time and space constraints relative to theoretical maximums, examine the losses and wastes associated with these, and to reduce these losses. All scale innovations reduce the time required per unit production, which reduces costs because time equals money. Examples include:

➤ In agriculture, horse-drawn plows led to tractors, which led to very large modern combines (45 feet across). Larger-equipment scale increases allow more work to get done per unit of time, directly reducing costs.

➤ The wind industry discovered 20 years ago that turbine output power is proportional to the rotor diameter cubed. This higher scale innovation lead to rotors longer than 100 feet, which increased performance dramatically, reducing costs on a dollar per kWh basis so much that wind power now competes with coal-generated electricity.

➤ Dramatic transistor size reductions drove the computer revolution, leading to exponentially reduced costs and increased performance.

This phenomenon is called Moore's law, and scientists predict that we will reach its limits in the next decade.

➤ Concentration innovations are another example of reducing the size of active components:

    o Transmission of electricity originally used direct current. It suffered from resistance losses, and therefore could not be used over long distances. Using higher voltages allows us to transmit much larger amounts of power (voltage x current)—to concentrate it—because the wire's resistance losses are governed by the current (per Ohm's law, Resistance = Power/Current^2). Introduction of alternating-current transformers allowed us to inexpensively "step up" and "step down" voltages, giving rise to modern transmission systems. Transmission voltages started at about 110 volts[183] in the late 1880s in New York, and then eventually increased to over a million volts[184], reducing current and resistance losses dramatically. This concentrated the power moving through transmission lines over 9,000 fold. As a result, transmission distances increased to what we use today, with the longest transmission line of over 1,400 miles transmitting into the interior of Brazil from the coast. DC systems, with fewer losses and wires, are used for long distance transmission networks, while AC systems, with less expensive step-up and step-down, are used for medium and short distances.

    o A second example is the wind turbine, which concentrates very dilute wind energy. First, energy is concentrated by using a fan blade rotating on a shaft. The fan blade area to total swept area ratio is usually 1:20, or a 20 times concentration factor. The energy in the blade(s) is then concentrated further to a large central shaft, at a 1:362 ratio. Gearing concentrates energy further down another 1:20. Overall, a greater than 100,000 times concentration factor applies to the input wind energy[185] relative to the output electricity wire size. This reduces cost dramatically compared to solar, and is a primary reason the large-rotor concept is the dominant wind-energy design today, and why wind energy is competitive with coal generated electricity.

➢ Another reduced-size innovation is to lower the number of active parts in a design, as espoused by "Design for Assembly" techniques[186]. For example, hydropower technology initially had separate pumps and turbines. When these two functions were combined into a single reversible pump-turbine, capital costs were cut in half. Further concentration is provided by water, which is 800 times denser than air —a cubic foot of moving water contains 800 times more energy than an equivalent breeze. Using a dam, with a relatively small turbine inlet vs. the entire dam area, concentrates the energy further. As a result, hydropower is the cheapest form of power available today.

➢ In factories, high volume production processes such as injection molding have low per piece costs compared to lower volume processes, such as machining. If products are sold in high volume (i.e. at larger scale), costs can be reduced. High volume production processes have higher initial costs (for molds, etc.) and lower cycle times, enabling them to produce more product per unit time.

➢ Automation is also a scale improvement. The most common type of automation is not full replacement of human labor, but machine assistance. Headsets and order sequencing computers at McDonald's are good examples. If a machine or software can assist humans with their labor, enabling more to be accomplished per human labor hour, this time savings is equivalent to larger equipment scale. More work can be done per unit of time, allowing higher production volumes to reduce costs. Full replacement automation (much less common given machines' poor but improving pattern recognition capabilities, inflexibility, and low intelligence) can provide savings when both labor costs are high and production volumes are high, as we have seen for dedicated factory automation equipment.

Innovations of scale—our ability to produce lots of stuff —have driven our exponential economic growth since the Industrial Revolution. Because this economic growth depends almost wholly on using fossil fuels, which are concentrated stores of "free" energy, our emissions have similarly grown.

Paradigm Shifts

Paradigm shifts are innovations that change systems. For example, coal, oil, natural gas, electricity, or geothermal have successively replaced wood fuel for space heating. Each paradigm has its own performance and cost limitations, to which the above lean engineering principles of efficiency and scale can be applied. For example, for home heating:

(1) Coal is a more energy dense fuel than wood, which reduces transportation, collection, and storage costs for heating a home.
(2) Natural gas burns more cleanly and when piped in has even lower transportation costs. Propane or home heating oil in nearby storage tanks can be used when natural gas piping is not available.
(3) Electricity-fueled resistance heaters are 99% efficient compared to today's coal, oil, or gas 83%-95% efficient home furnaces, but resistance heaters use electricity that is inefficiently generated by a central power station (for example, if coal-fired, incoming electricity is 30% efficient; if natural-gas fired, about 60%).
(4) Heat pumps use a reverse refrigeration cycle to heat spaces at triple the efficiency of resistance heaters in temperate climates, but heat pumps lose efficiency as the outside temperature decreases, becoming equivalent to resistance heaters at 0°F.
(5) Geothermal heat pumps use the Earth as a heat source for this reverse refrigeration cycle. The Earth's constant warmer subsoil temperature substitutes for colder outside air, allowing 3 times the efficiency.

In general, the higher efficiency options in this list are costlier, as heat exchanger size and cost increases with efficiency.

As we shift paradigms —from horses to bicycles to automobiles to jet-packs, or from ships to railroads to trucking —the new paradigm has reduced losses compared to the old, to allow higher scale and lower costs. In some cases, a particular paradigm will fit a particular market niche or application/geography. For example, heat pumps are efficient in temperate climates but should not be used in northern Maine, and shipping by sea is inexpensive but slow, while air shipping is more expensive and much faster.

The Economic Law of Pollution —an empirical collection of our experience reducing pollution over the last 100 years —states that the least expensive way to reduce pollution is through efficiency gains (i.e. efficiency and scale innovations). The second least expensive way is through a paradigm shift, to shift to less-polluting resources. Currently, our society is wholly dependent on fossil fuels, and substitution is not quite yet technically or economically feasible. To make alternatives viable, lean engineering shows that we need to focus on reducing losses and waste relative to the ideal.

The general process by which we have achieved continuous improvements over the modern age is shown in Figure 29. This process has been the primary engine (apart from population growth) of most of our economic growth since the industrial revolution, including the massive increase in our living standards.

Continuous Improvement

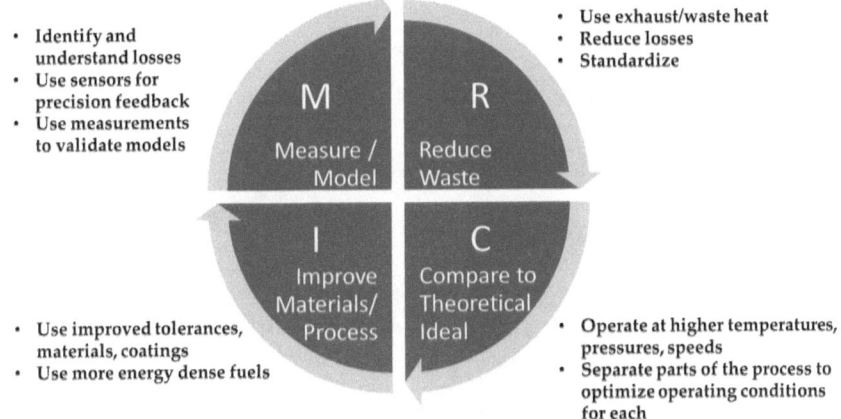

- Identify and understand losses
- Use sensors for precision feedback
- Use measurements to validate models

- Use exhaust/waste heat
- Reduce losses
- Standardize

M Measure / Model

R Reduce Waste

I Improve Materials/ Process

C Compare to Theoretical Ideal

- Use improved tolerances, materials, coatings
- Use more energy dense fuels

- Operate at higher temperatures, pressures, speeds
- Separate parts of the process to optimize operating conditions for each

Figure 29. Historical Technology Improvement Process (MRCI)

Economic Law of Pollution Applied

The rest of this chapter applies the Economic Law of Pollution to our top emissions sources, in 80:20 Pareto fashion, to determine what short-term actions we can take to reduce fossil fuel usage, saving on long-term energy costs and spurring economic growth, while reducing emissions to solve global warming. I delineate how close to the ideal we are for each source to show its potential to reduce emissions via efficiency or scaling; and I consider other paradigms and their relative potential.

Top U.S. Emissions Sources and Trends

A key part of America's general strategy for pollution reduction is to use Pareto principles to find the lowest cost solutions. If 80% of the effects come from 20% of the causes, focus on these first. Current sources of CO_2 emissions in the United States are shown in Figure

30^{xvii}. Base load coal power plant emissions for electricity, tailpipe emissions from automobiles, and peak load emissions from natural gas power plants for electricity are the top three sources, with the first two sources comprising over half of our emissions. These top three sources will therefore be the primary focus of this chapter.

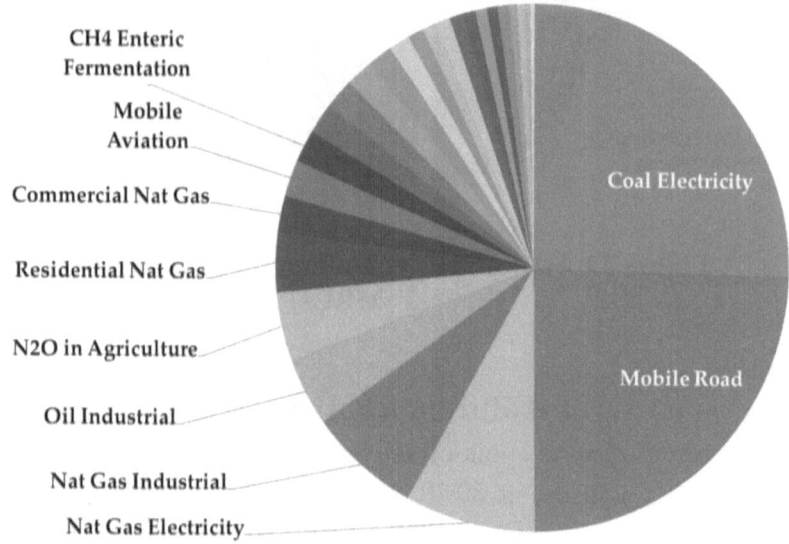

Figure 30. U.S. Greenhouse Gas Emission Sources in 2010[187]

Note, however, that the U.S. has sent much of its manufacturing CO_2 emissions to other countries over the last two decades. Therefore it is likely that industrial sources are severely underrepresented in Figure 30, if we account for all of the emissions that support our lifestyles. Because of the developing world's almost non-existent pollution controls, fluctuating trade flows, and varying development levels, it is difficult to properly account for these emissions.

---

xvii Note that sources are shown in million metric ton (designated Tg) CO_2 equivalents, as a number of gases (methane, etc.) have more powerful warming effects than CO_2, so they may have significant impacts at much lower concentrations.

For instance, cement emissions worldwide are relatively high (5% overall) due to construction in the developing world. In the U.S., where new construction activity is much lower, cement emissions are much lower. Appendix B, which analyzes sources 4 through 10, captures some of these effects. But because sources 4 to 10 each comprises only 6% or less of our emissions, reasonable savings levels of 10%-20% each comprises only about 1% savings each, which is too little to make an appreciable difference.

As the second largest CO_2 polluter in the world, overtaken by China in 2006, the U.S. emitted about 5.4 trillion metric tons of CO_2 in 2014, which is about 10% lower than U.S. emissions in 2005 (see below). The reduction in emissions was caused partly by the recession, which depressed economic activity, and partly by the switch from coal to gas fuel, as natural gas plants are more efficient and emit about half the CO_2 of coal plants[188]. Fracking increased the supply of natural gas many-fold, and this glut has led to historically low natural gas prices. As a result, utilities have been phasing out aging coal plants in favor of natural gas, reducing both their operating costs and emissions.

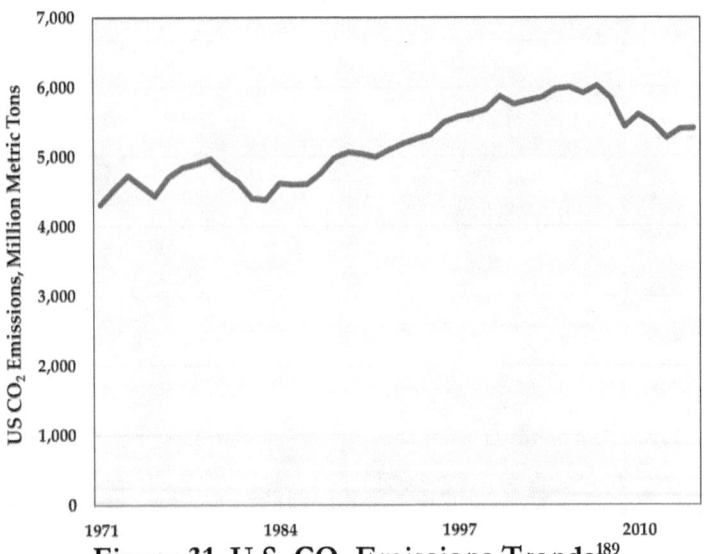

Figure 31. U.S. CO₂ Emissions Trends[189]

The US Energy Information Administration projects that this trend will continue, allowing the U.S. to meet the original Kyoto protocol requirements through the unintended consequence of technological progress, rather than through regulations.

Let us now consider how the Economic Law of Pollution applies to our top three emissions sources.

Strategy #1: Improve Efficiency or Utilization of Resources

Coal-fired Generation

The top CO_2 emissions source in the United States is coal-fired electricity power plants. Applying the 80:20 Pareto rule, it is clear that the most inexpensive and effective way to reduce overall U.S. fossil fuel emissions is to start with coal power plants by substituting other power sources or by sequestering their emissions. This is echoed by economist Nordhaus, James Hansen, and many others, and it led to the EPA's Clean Power Plan Proposed Rule, which de facto makes coal more expensive.

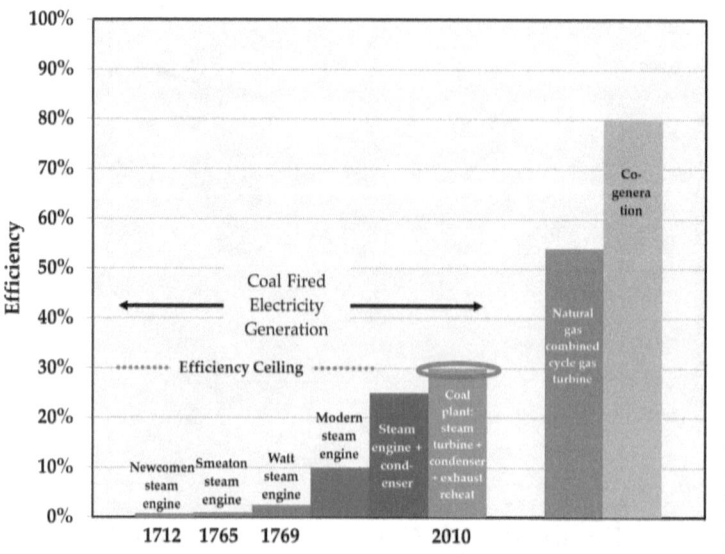

Figure 32. Steam/Gas Power Plant Efficiency[190]

As Figure 32 shows, progress in improving the efficiency of coal-fired generation has stagnated at a relatively low 30% (the oval) for two primary reasons: (1) As a heat engine, coal plants are limited by thermodynamic physics and Carnot's theorem to 70%-80% theoretical efficiency because firing temperatures are limited by materials of construction. (2) Low grade waste heat from burning coal cannot be easily economically converted into electricity.

However, it is possible to reduce this waste heat in at least two ways, as shown above the red dotted line.

(1) Co-generation schemes use the low grade waste heat to run factory processes or heat buildings. Factories and buildings need to be located near the power plant to economically reduce insulation losses.

(2) Combined cycle natural gas turbines burn gas at much higher temperatures, improving theoretical efficiency, and then use the waste heated gas in a secondary cycle to drive a turbine to obtain electricity. This strategy can be applied to coal; "coal gasification" schemes gasify coal, and then use a combined cycle process to obtain higher efficiency. However, it takes heat and pressure to gasify coal, so there are economic tradeoffs between this expense and the savings. So far these tradeoffs are not favorable, especially because the additional coal gasification equipment costs more.

In summary, both coal-fired and natural gas power plants have generally hit an efficiency ceiling.

Automotive

Figure 33 and Figure 34 show how the efficiency of an internal combustion engine (ICE) is also relatively low, approximately 20% for modern automotive engines. Diesel engines, with a larger expansion ratio, can approach 30% efficiency; and reach up to 45% for lower speed applications (e.g. boats).

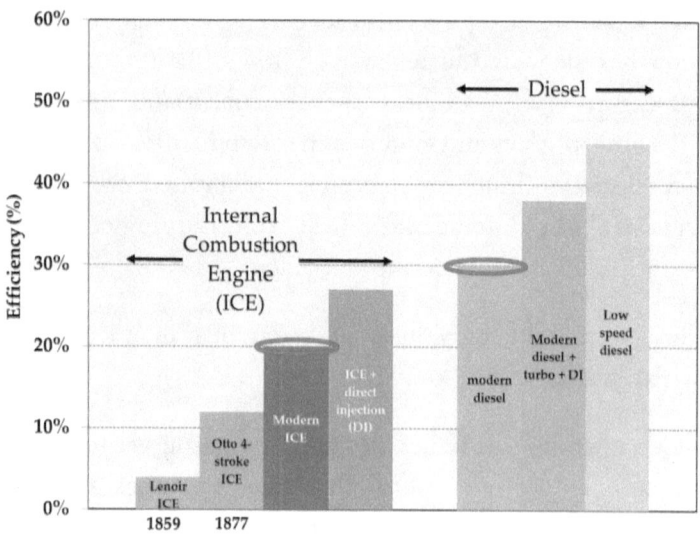

Figure 33. Internal Combustion Engine Efficiency

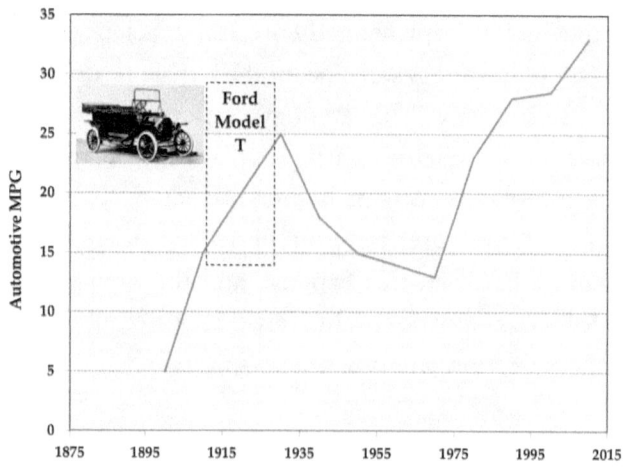

Figure 34. Automotive Efficiency History[191]

The figures also show how much engines have improved, and how mileage has changed. A large increase in CAFE fuel standards in the

1970s in response to the OPEC oil crisis doubled automobile efficiency, back to what it was in the era of the Model T[xviii].

With electric motors at 95-98% efficiency[xix], several hybrids on the road, and recent changes in CAFE standards, clearly we have not reached the limit of theoretical efficiency for automobiles.

Gas-fired Electricity

Figure 32 shows combined cycle natural gas plants with relatively high efficiency, about 55%. In a combined cycle gas plant, the hot gases from burning the natural gas at higher temperatures are used to spin a gas turbine and generate electricity, thereby using some of the waste heat to generate electricity. This efficiency difference, combined with low natural gas prices, has led to coal-to-natural-gas fuel switching that will reduce our emissions to Kyoto levels, as noted above.

To make natural gas more efficient and further reduce our emissions would require the use of cogeneration (i.e. CHP, combined heat and power). In cogeneration, the waste heat from a thermal power plant warms water or air that is then used for space heating or cooling. The source and load must be located close to each other, as insulation losses can otherwise be high. Loads can include hospitals, prison, manufacturing processes in factories, swimming pools, airports, hotels, apartment blocks, etc.

In addition, there are tradeoffs between central power generation and distributed power generation, which uses many smaller power sources more widely spread out to closely match source and load. Distributed sources reduce transmission losses, but also tend to be less efficient than central sources. As a result, there are a number of legal and other

---

[xviii] Note, though, that the Model T is more equivalent to a modern golf cart, and top speeds were low, so mpg is not strictly comparable.

[xix] However, strictly speaking, the efficiency of the source of the electricity utilized varies widely, depending on the fuel mix used to produce the power, even though the efficiency of transformation of electricity to motion is very high. Nuclear is >90% efficient, natural gas ~55%, coal ~30%, etc.

barriers to CHP[192] in the U.S. because of our widespread use of central generation. The most economical places for CHP are where thermal loads are concentrated and relatively high (i.e. near factories, on university campuses, etc.)

Strategy #2: Eliminate Pollution at the Source

Eliminating CO_2 pollution at the source requires a few changes to our fossil-fuel based economy. These include:

a) Switch from coal and gas to renewable power sources (hydro, wind, solar, biomass, nuclear, tidal) that emit no CO_2 in operation

b) In transportation, switch to electric or fuel cell[193] cars, wind-powered boats, and bio-fueled or electric airplanes

Power Generation

The history of switching to power generating fuels that emit less or zero CO_2 has shown that the market prefers to switch only when it makes economic sense, and when the resource is readily available. Let us now consider the potential options for switching from coal.

Natural Gas Substitution

Natural gas is less expensive than coal, as U.S. utilities have demonstrated in the last few years. They have substituted natural gas plants for retired coal plants over the last decade, despite natural gas plants' higher initial capital cost. If cogeneration is used to increase efficiency (by clustering thermal loads near power generation plants), higher plant efficiencies could reduce the cost of this option even more.

However, there is some controversy over using natural gas as a "bridge" fuel from coal to renewables because of the fear that stray methane emissions from drilling, tapping, transporting, and using natural gas are underestimated. These fears increase when one considers that methane is 20 times more effective as a greenhouse gas than CO_2, which means that very low emissions rates can have high impacts. While the industry has economic incentives to reduce these emissions, in some cases they don't[194], so it makes sense to measure stray methane emissions. The EPA issued regulations in 2012, which continue to be modified. In addition, some environmentalists decry substituting natural gas for coal because it doesn't really address the root problem—somewhat similar to substituting methadone for heroin, it reduces emissions, but doesn't address the underlying addiction. However, given the status of alternatives and the proven effectiveness of this substitution, natural gas can and likely will serve as an imperfect "bridge" fuel in the coming decades. It may not work as well in China because natural gas is deeper underground, which increases costs relative to coal, and because they lack pipelines. But it will work for the United States in the near term.

Micro-turbines (small locally operated natural gas turbines) can also be used for power generation. But, due to their small size, micro-turbines are much less efficient than larger power plants. The cost of higher efficiency combined cycle equipment is higher at small scales, even when accounting for reduced transmission losses of 4-8%. Higher efficiency gas turbines also use expensive high temperature materials. These relatively high fixed costs imply that there may be a minimum scale for gas turbines to be practical; in addition, the production scale of micro-turbines in general is low which increases costs. High costs have led to minuscule market share, and micro-turbines are used generally for CHP applications.

Renewables Substitution

Hydro has essentially penetrated into where suitable waterways are available, as extremely long project lifetimes lead to the lowest electricity costs available, once the high initial cost is incurred. While there is further potential for micro-hydro projects, the economics of these are less attractive as relatively high environmental/project costs are spread out over lower power outputs.

Geothermal power has relatively high capital costs, and is similar to hydropower in that the locations where it is available are limited. Suitable locations are found where high temperature sources are near the surface (i.e. near hot springs, etc.)

Wind competes[195] with coal without subsidies, but not with lower cost natural gas, and is highly variable. It also becomes less competitive when transmission costs are factored in, as many of the best wind sites are far from loads (i.e. cities). In addition, it is not a solution in the Southeast as there the wind dies off in the summer, as shown in Figure 35.

SUMMER

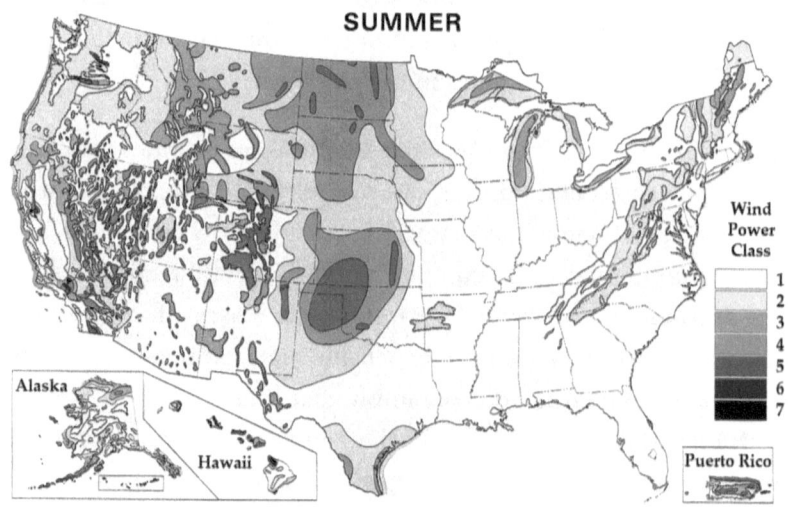

Figure 35. U.S. Summer Wind Lull[196]

Solar is not competitive without subsidies in the U.S., despite huge cost and price drops in recent years. However, it has reached grid parity in a few locations, especially during times of peak demand on hot summer days.

Wind and solar variability causes problems because the wind blows at night when the power is needed less and drops off in the summer in some areas; solar works well during sunny days, but doesn't meet 5-8 p.m. residential peak needs, or needs on very cloudy days, or needs during the night.

The electricity grid can accept only so much power from variable sources. As a result, it will require economically viable energy storage to break through these limits. At the moment, we use fossil fuel, geothermal, hydro, and nuclear sources to provide backup power during times when variable sources are not available.

This variability problem is reduced when one geographic area can compensate for another—i.e. when one location is windy while another's winds have died down, or when one location is cloudy while another is sunny. If transmission exists between these areas, variability and the need for backup power decrease.

Wind and solar are also complementary, as each tends to produce power when the other doesn't. Denmark has achieved greater than 25% penetration with wind power through close transmission connections with its neighbors, and through thermal storage in its large combined heat and power (CHP) plants. It has a goal of reaching 50% wind by 2020.

California achieved 20% penetration of a combination of renewable sources (wind, solar, hydro, geothermal, and biomass) in 2009, and is on track to achieve 33% penetration by 2020. Along with high levels of transmission to surrounding areas, storage to meet these goals can potentially be provided by hydro (pumped storage) and thermal (CHP, hot water heaters, concentrating solar power (CSP)[xx]) sources.

Based on these examples, from a technical perspective, it appears feasible to achieve 30%-40%[197] penetration of variable sources throughout the U.S., especially if national transmission grids are upgraded to allow better sharing of variable sources. Economically,

---

[xx] Some trough plants provide thermal storage as they use the sun to heat a heat transfer fluid; this heat is then used to generate steam and then electricity.

wind is as inexpensive as coal on a new power plant basis, but may not be able to compete with coal when incumbent capital advantages are factored in (i.e. plenty of coal plants are already paid for); and solar energy can be competitive in some sunny locations against retail electricity rates (even without 30% investment tax credit subsidies). Therefore renewables can be economically substituted for coal plants as they retire, up to the technical limit of 30%-40%. But beyond this limit, economical storage technologies that don't yet exist are needed.

Nuclear Substitution

In the United States, there have been no nuclear plants built since 1974 because nuclear activists have driven construction costs much higher, illustrating not-in-my-backyard (NIMBY)-ism writ large. In addition, the Three Mile Island accident, the Fukushima plant accident in Japan, and the Chernobyl accident in Europe have demonstrated how relatively unsafe this technology is.

In the U.S., there have been at least 56 nuclear reactor accidents since 1952[198], including numerous instances of leakage into groundwater supplies. On-site storage of nuclear waste, and nuclear waste in general, remain serious issues that have not been resolved, and "nine states have explicit moratoria on new nuclear power until a storage solution emerges."[199]

But James Hansen and others believe that nuclear power is the only commercially viable near-term substitute for fossil fuels. I agree with Hansen that breeder/next generation nuclear reactors, if proven at scale, can potentially use our current nuclear waste as a CO_2-free fuel source, and I fully support research into this cleanup method as a possible partial solution to the nuclear waste storage problem. But it is clear that nuclear fission power generation will not be able to reduce our CO_2 emissions safely, because accident liability risks are so large that private insurers are unwilling to shoulder these risks for a mature industry.

92

Nuclear fusion research has not yet produced break-even (self-sustaining) controlled fusion reactions, so is currently unfeasible. It will also produce nuclear waste (albeit much less than fission, and lasting only a few hundred years rather than thousands).

Summary

In summary, renewables should be used to replace the current fleet of aging coal plants, up to 30%-40% penetration. Higher variability sources should be addressed through energy storage (via combined heat and power (CHP), pumped hydro, concentrating solar power (CSP), and hot water heaters), higher capacity national transmission networks, and natural gas backup. Beyond this limit, lower cost renewables and energy storage will be needed, so we should pursue accelerate research into these solutions.

Transportation

As can be seen in the 2014 and 2015 U.S. government projections[xxi] below, oil prices of $20 a barrel during the last two centuries of economic growth are likely gone forever. Prices are projected to dip to $50 a barrel in the near term as new oil domestic supplies are used up, and then prices will likely increase to over $80 a barrel for the foreseeable future. "Peak Oil"[xxii] has been delayed by the use of fracking techniques, which is remarkable, but it is clear both worldwide and in the U.S. that most of the cheap sources of oil— i.e. conventional resources —are dwindling. What is left are non-conventional sources, such as tar oil, oil sands, tight oil, etc., with higher extraction costs, meaning prices will rise over the long term.

[xxi] The track record of U.S. government projections is relatively low, with prices usually underestimated. Figure 36 may be conservative.
[xxii] This term refers not to running out of oil altogether, but to running out of *cheap* oil.

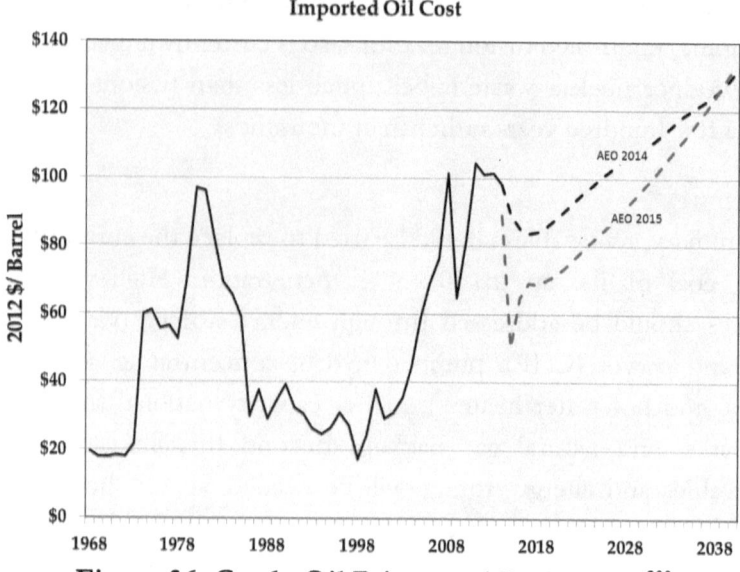

Imported Oil Cost

AEO 2014

AEO 2015

Figure 36. Crude Oil Prices and Projections[200]

As the price of oil (and derivative gasoline, plastics, etc.)[xxiii] rises, transportation ideas that shift to non-fossil or more efficient sources are inexpensive ways to reduce CO_2 emissions. Despite a high initial price, the fuel cost savings will likely pay for emissions reductions.

Diesel

As mentioned in the section on energy cost myopia, diesel vehicles are generally more efficient than gasoline but cost a little more, with a three year payback. So why don't Americans drive diesel cars? In response to gas price spikes in the 1970s, Detroit released diesel models, which had problems. They were highly unreliable, underpowered, noisy, smelly, and didn't start in the cold; the recent Volkswagen EPA emissions scandal reinforced diesels' poor reputation in the U.S.

These quality issues have been resolved (at least in foreign car markets), which means that the market barrier is artificial. Diesel prices, which had been 4% less than gasoline prior to 2004, have since become 6%

[xxiii] Higher than the traditional $20 a barrel.

higher due to a combination of a 6 cents per gallon federal tax (another political market distortion), U.S. refinery practice[201], and high foreign diesel demand. This differential tax should be repealed to eliminate this market distortion.

Natural Gas Vehicles

Even though fracking is increasing natural gas supplies, and prices are low, at 2 \$/gallon equivalent, natural gas cars and trucks make sense only for rental car fleets and buses because conversion costs are \$6,500-\$12,000. These costs include a fuel tank, regulator, controls, and a compressor to switch between .5 psi at home and 3,600 psi in the vehicle; fuel fill up compression times are measured in hours. Such high prices give unreasonable paybacks of 6-plus years for the typical U.S. family car, and may move higher if gasoline prices drop. As a result, 12-15% of public transit buses operate on natural gas. Another major market impediment is the limited number of natural gas fueling stations—there are only about 900 in the country.

Electric Vehicles

Currently fossil fuels, in the form of gasoline and jet fuel, are best suited for transportation because their energy density is so high[202], reducing vehicle weight and increasing range between fill-ups. The energy density of batteries, even lithium ion, is less than $1/50^{th}$ that of gasoline.

The emissions benefits of electric cars also depends on how much carbon is used to produce the electricity that drives the vehicle[xxiv]; if coal produces the electricity, then electric cars can make emissions much worse. Smart connection of electric car batteries to the grid could potentially provide some energy storage to allow higher levels of

---

[xxiv] China, for instance, is exploring electric car deployment heavily. But its electricity system is driven primarily by coal, so this will likely lead to more carbon emissions than if gasoline were burned directly (due to transmission losses and relatively inefficient coal power plants compared to natural gas).

variable renewable power integration[xxv]. Hybrid electric/fossil fuel engines can work well to increase gas and/or air mileage (as the Dreamliner and the Prius demonstrate), but current 100%-electric car ranges are almost exactly the same as twenty years ago, except for the recent Tesla Roadster[203], because battery innovations have stagnated for decades.

The jury is out on how well the Tesla technology will work at larger scale. Issues include lithium environmental mining limitations, weight capacities, high battery costs even by the GigaWatt (GW)-sized factory, and limited battery lifetimes. If battery costs for Tesla's GW plant turn out to be low, or lifetimes long, then the high end Tesla S can achieve cost parity, but at a much higher up-front cost (for the battery). At the moment, however, a 100% electric vehicle is a luxury concept that is not readily scale-able for everyone. Hybrids make more sense, but are still less than 4% of the market.

Finally, a 100% electric plane is not feasible, except in very low-speed/glider configurations, as power to weight ratios for batteries vs. jet fuel are insufficient.

Wind Vehicles

The substitution of wind power as a primary mover works well at sea, but is typically much slower (for instance, sailboats tack back and forth to go against the wind at much slower speeds than a motor-powered boat). Wind is also not available 60%-80% of the time.

Light-weighting

With its 787 Dreamliner, Boeing went to an all-electric control architecture and changed to composite bodies rather than all

---

[xxv] Conversely, a high level of electric cars and charging stations may impact grid reliability, especially if charging is done quickly, however, with time and innovation such load balancing issues may be resolved.

aluminum, reducing plane weight by 20%. Automotive and truck manufacturers could do the same[204].

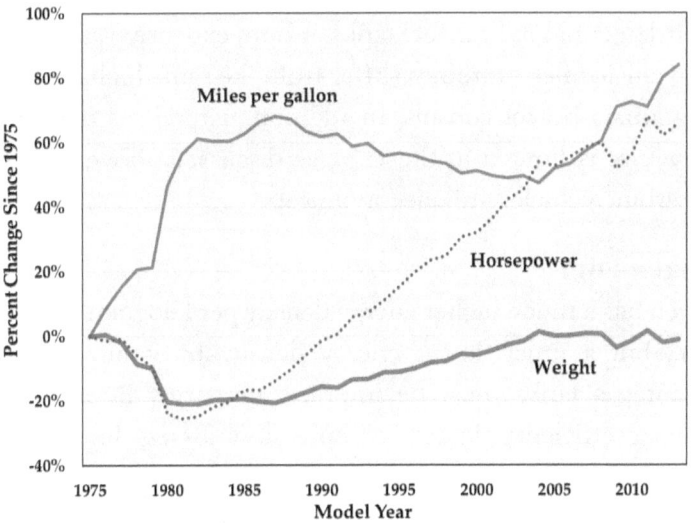

Figure 37. Automotive Vehicle Trends, 1975-2014[205]

After the first gasoline spike in the mid-1970s, vehicle weight reduced by 20%, but has since crept back up to an average of 4,000 pounds per vehicle. Every 100 pounds of weight reduced improves miles per gallon by 1-2%.

Aerodynamics Improvements

Airplane aerodynamics continue to improve, as airfoils, flow control, composite structures, and propulsion improvements increase airplane fuel efficiency.

Automobiles and trucks do not look like streamlined submarines or fish. As a result, their aerodynamic drag coefficients are relatively high, with room for improvement. One of Tesla's innovations to extend range, in addition to using a large battery, has been to improve the drag coefficient (.24 vs. .35 for a normal car) to reduce aerodynamic losses.

Micro-turbines

Micro-turbines can also potentially be used to generate electricity to improve electric vehicle range. The addition of a mini power plant, and relatively larger natural gas fuel tanks, is both expensive and heavy, and electric vehicle micro-turbines suffer from the same high cost issues as their stationary-power cousins. In addition, natural gas fueling station infrastructure is limited in the U.S., as discussed above. As a result, micro-turbine vehicles are uneconomical.

Hydrogen Fuel

Hydrogen has a much higher energy density per kilogram compared to gasoline, but a much lower energy density by volume. Therefore equivalent gas tanks must be impractically large. Because of 50% electrolyzer efficiency losses, compression losses, higher pressure vessel tank costs, fuel cell high platinum costs and low *system* efficiency, hydrogen fuels are uneconomical compared to gasoline, even at large scale. In addition, the least expensive source of H_2 is natural gas, so alternative methods of H_2 production will be needed if it is to replace fossil fuels[206].

Biofuels

Whether we burn plants that grew millions of years ago, or plants that grew just recently, CO2 is released. Recently grown biofuels put about the same amount of CO_2 into the air as oil-based fuels. However, if one looks at the complete lifecycle of biofuels, including the CO_2 generated in harvesting and processing, the CO_2 generated upon burning, and the CO_2 ingested during plant growth, the CO_2 ingested during plant growth can mostly compensate for its other emissions[207] —which is much better than fossil fuels.

Yet biofuels suffer from issues involving harvesting, energy density, land use, and equipment costs. Bio-feed stocks are generally available only a few times a year, and the costs to collect, densify, and process them are high. Biodiesel costs are greater than $4 a gallon as of 2014[208].

From an energy density perspective, one disadvantage of bio-feed stocks relative to fossil fuel sources is lack of a free 2:1 compression factor, as fossil fuels were compressed into a more densified form over millions of years. This reduces fossil fuel shipping and processing costs compared to freshly grown trees/biomass.

Biofuels also impact land use, and there has been great debate recently if land use changes induced by higher bio-fuel use paradoxically increase greenhouse gas emissions enough to offset the benefits of biofuel use in the first place[209].

This is fundamentally driven by relatively low "feedstock energy density per square foot of feedstock growth area" considerations. Plants generally need space to grow, and biofuel crops (grasses, rapeseed, corn, etc.) tend to have a low level of bio-feed-stock mass per acre planted. This is the primary reason why U.S. ethanol policy is an abject failure[210], and should be jettisoned. Sugarcane energy density is twice that of corn, so Brazil's ethanol scheme is better than the U.S. corn scheme, but is not practical to copy as we do not have a tropical climate.

Summary

In conclusion, there are *no alternatives* to the use of fossil fuels for transportation that can compete with gasoline and jet fuel and offer the same convenience and utility. Without further breakthroughs, full substitution is not a reasonable option. However, low hanging fruit of aerodynamic improvements, weight reductions, and electric-hybridization can significantly improve fuel miles per gallon.

If the market is activated, as described in the previous chapter, consumers will consider fuel efficiency, safety, cost, aesthetics, reliability, etc., as they purchase new vehicles. But because of our energy cost myopia, we will likely not value fuel efficiency highly enough[211], making inexpensive higher efficiency options look expensive. As a result, I endorse the need for much higher CAFE standards than we have— we should simply require twice our current fuel efficiency, and let the market decide which approaches make the most sense(i.e. light-weighting, aerodynamic improvements, hybrid electrics, etc.) Higher vehicle costs will be paid for by fuel savings. Doing so will reduce our total fuel demand and thereby help reduce fuel prices, partially compensating for the ~5% GDP impact that increased fossil fuel prices have on our economy.

Current CAFÉ standards are this high, proposing to double vehicle miles per gallon. However, differences between real-world and laboratory test conditions, CAFE credit trading, and other fleet loopholes will reduce the *actual* fuel economy achieved. These loopholes should be closed to allow our economy to grow.

Strategy #3: Capture Pollutants When Emitted

After Strategy #1 (improving efficiency) and #2 (source substitution), the third strategy in the Economic Law of Pollution is to capture pollutants when emitted. Pollution controls for cars and power plants capture or transform pollutants at the point of emission. One example is catalytic converters, which combust CO to CO_2, unburned hydrocarbons to CO_2, or reduce NO_x to N_2; each of these reactions sharply reduces CO, hydrocarbon, and NO_x emissions. Another example is wet scrubbing of SO_2 at the exhaust to coal plants. Typically[212] alkaline absorbents are used to absorb SO_2, and this material is landfilled. The same technique, but applied to CO_2 rather than SO_2, is called carbon sequestration, and consists of three processes: carbon capture, transport, and storage.

Carbon Capture

CO_2 in the atmosphere diffuses to the maximum extent possible, and it takes a minimum amount of work and energy to reverse this diffusion process of about 400 kJ/mole[213]. This is illustrated below in a Sherwood plot, which shows that the cost of separating a substance from a mixture rises as the concentration of the substance being extracted falls. However, while the energy required to capture carbon at the exhaust pipe rather than to separate it from the air is lower ($50-$100 a kg vs. $100-$1,000 a kg), there appears to be a minimum cost threshold as concentration increases (i.e. the curve flattens out).

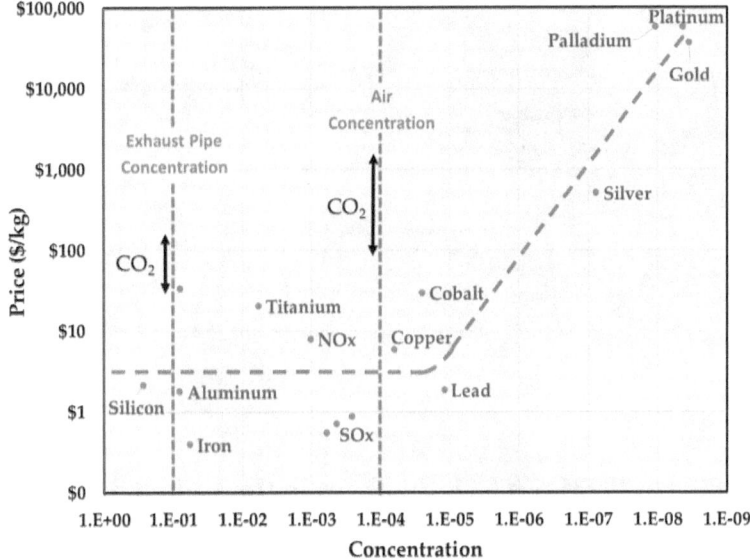

Figure 38. Sherwood Plot[214]

Chemical capture methods use chemical absorbents such as NaOH or amines. The issue is not capturing the CO_2 in the first place—this is energetically favored for many substances —but in getting these substances to cost effectively release the CO_2 once it is captured. Global Research Technologies is working on a NaOH absorbent plus resin system that binds CO_2 when dry, and releases it when it is wet. Global Thermostat is working on an amine system that binds CO_2 easily, and uses low grade heat to release it. However, it is still an open question if these processes can be made inexpensive enough to deploy.

Transport
CO_2 pipelines already exist in the United States, as shown in Figure 39, covering about 4,000 miles; this is a small fraction of the natural gas pipe infrastructure of the country. The technology is well understood and technically feasible, as CO_2 is generally stored and transported as

a stored liquid for enhanced oil recovery (EOR) operations and projects, and costs \$850,000-\$1.5 million a mile[215].

Figure 39. Major CO_2 Pipelines in the United States[216]

Storage

Once we've captured the CO_2, where can we safely put it so it will not get back into the atmosphere? While there are many possibilities, let us consider the leading ideas of storing CO_2 (a) in underground reservoirs, (b) under the ocean, (c) within minerals, (d) in soils, and (e) in forests.

Underground Reservoir Storage

The oil and gas industry creates holes in the Earth as its bores down to extract and frack these resources. Carbon could be stored deep underground as liquid CO_2 in these same reservoirs; this is termed classical carbon capture and storage (CCS). In fact, 65 million tons[217] of liquid CO_2 is already used annually in the U.S. for enhanced oil recovery (EOR) operations to help obtain more oil, and one of EOR's economic limitations is the lack of inexpensive sources of liquid CO_2.

However, while this storage may be effective for the most part, there is no way to ensure that the sequestered CO_2 would not be subsequently released—for example from a cap/sealing failure, an earthquake, tsunami, or future drilling operation. Because CO_2 is heavier than air, releases suffocate nearby people and trees[218]. As a result, not-in-my-backyard attitudes are a big issue at storage sites near people. This is similar to the situation with nuclear waste disposal, but if we were to sequester CO_2, the number of tons and volume would be much, much larger, with a subsequent higher potential for release. So we could spend a lot of money sequestering CO_2, and not know whether it would be effective long term. But these dangers may be exaggerated by opponents of CCS —caps can be made of stone/concrete to last, earthquakes and tsunamis occur in known locations (i.e. the Ring of Fire), monitoring can be set up, and future drilling operations will be able to predict the CO_2's location as we get better at underground sensing.

Another potential unintended consequence of underground storage would occur if the CO_2 leaked into a rock formation that can mineralize. In that case, the mineralized rock's volume would double, with the potential to create earthquakes. We may be able to mitigate this possibility through our knowledge of the storage reservoir's geology.

Under-ocean Storage
Similarly, one could sequester liquid CO_2 by pumping it to the bottom of the ocean. Because liquid CO_2 is denser than water, it would likely be safely stored, except that the CO_2 would inevitably mix with water to form carbonic acid, and therefore lead to higher levels of ocean acidification, to the likely detriment of marine ecosystems.

Mineral Storage
Finally, it is possible to convert CO_2 into carbonate minerals. Typical schemes contemplate pumping CO_2 and water into ultramafic rock formations (such as peridotite and serpentinite). While favorable

thermodynamically, mineralization occurs slowly. Additional chemicals, water, heat, or higher surface area could make the reaction proceed more quickly, but would raise costs. In addition, when mineralized, the volume of the mineral is different from its components, which can lead to high pressures and earthquakes, reactant starvation, etc. So far, this route does not appear to be practical.

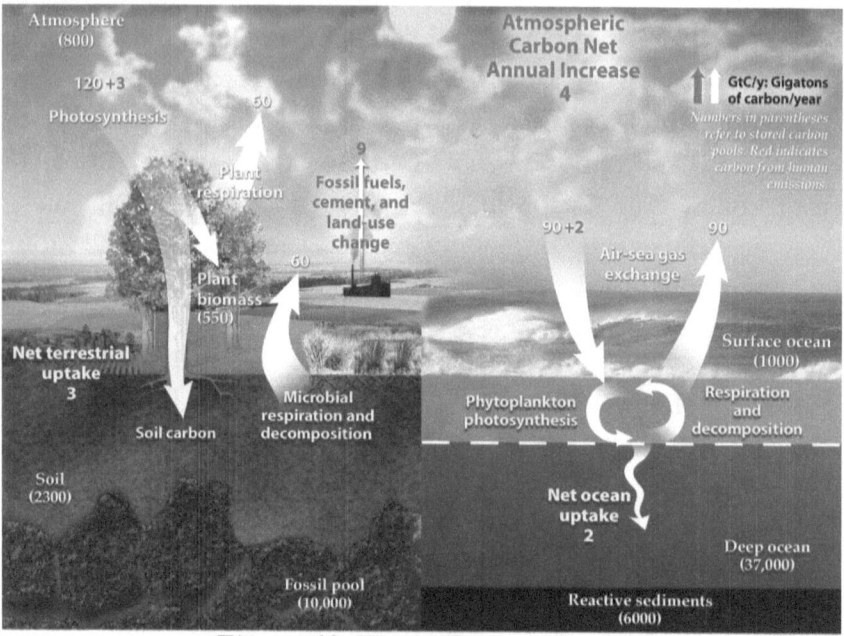

Figure 40. World Carbon Cycle
Yellow = natural flux; red = human flux; white = stored carbon.[219]

Soil Storage/Compensation

In plants, CO_2 is captured in complex organic molecules, which are fragile and subsequently break down to release CO_2 (i.e. the 60 billion tons of carbon shown as microbial respiration and decomposition in Figure 40). This respiration can be seen in the Keeling curve in Figure 41, which measures Earth's breathing, the annual global CO_2 uptake and release from plants. Most vegetation is in the Northern hemisphere, where 68% of Earth's land is located[220]. CO_2 levels decrease from northern spring onwards as new plant growth takes carbon dioxide out of the atmosphere through photosynthesis and rises again in the northern fall as plants and leaves die off and decay to release the gas back into the atmosphere[221].

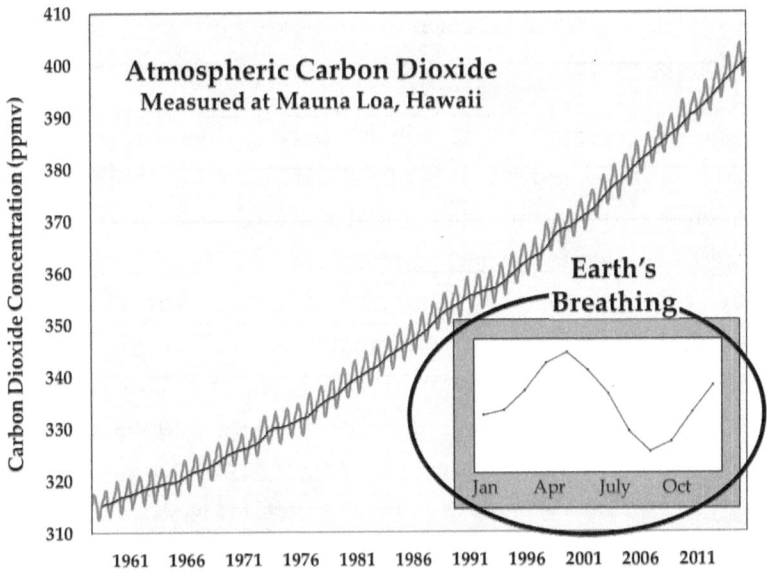

Figure 41. Keeling Curve, Earth's Seasonal Breathing[222].

One method of storing more carbon in the soil is to look for forms of plant matter that do not easily break down, which leaves their carbon content in the soil and reduces microbial respiration and decomposition. As one example, humus is relatively stable, and can help increase the carbon content per acre in soil. Can we use our fossil fuel emissions to create enough humus or "Terra preta" (another stable carbon-heavy fertile soil found in the Amazon), adding CO_2 permanently to the soil rather than releasing it into the air? As shown in Figure 41, decomposing plants release CO_2—can we stop this process, and somehow hold this otherwise released CO_2 within the soil?

Humus requires fungi, moisture, nitrogen, and calcium for its formation[223], but we have no methods of otherwise stopping plant decomposition or using CO_2 coal plant emissions to assist this process. Similarly, to form "Terra preta", one starts by adding low temperature charcoal, bones, fish, pottery shards, ash, manure and unknown micro-organisms into soil, but precisely how to create it is still unknown and a subject of research.

We could sequester CO_2 in soil by burying biochar. To produce biochar, one takes bio-waste (wood debris, manure, food processing waste, etc.) that would otherwise decompose and heats it in the absence of oxygen in a sealed furnace at 350-700C. Heat is released, sustaining a pyrolysis reaction, and charcoal, syngas, and oil is produced; the latter two can be burned to produce electricity and fuel the process. When the about 50% biochar output is buried in soil, this sequesters the carbon within the soil for long periods of time; in addition, soil fertility increases, allowing reduced fertilizer use. The porosity in biochar also helps soil retain water. However, the long-term advantages and disadvantages of biochar application have not yet been fully established.

Cost estimates for biochar range from $50-$2,000 a metric ton[224] (U.S. Biochar Initiative, Shackley[225] , Galinato[226]), with a sort-of consensus between $300 and $500. Key uncertainties include the production conditions, feedstock types, and harvesting conditions. Harvesting bio-feedstock from too far away increases transportation costs, as does transporting it back to farms for use. Because feedstocks are only harvested a few times each year, equipment is not fully utilized year-round, wasting capital. In general, however, unless the long-term benefits of biochar usage become better established and more measurable, the initial cost appears relatively high.

A number of other land management practices can reduce CO_2 release from soils. These include (1) reducing/no tillage practices[227] (tillage exposes soil to air and increases oxygen reactions within the soil); (2)

erosion control (contour plowing, terracing); (3) addition of organic amendments (compost, manure, crop residues); (4) reducing fertilizer usage[228]; and (5) use of cover crops. However, it can take decades for these practices to significantly impact soil carbon levels; and soil organic carbon levels are difficult to measure[229], so while these are relatively inexpensive, their effectiveness may be limited.

Forest Storage/Compensation

Approximately 33% of human carbon dioxide emissions is from clear-cutting[xxvi] and burning of forests worldwide, especially in the Amazon[230]. Table 4 below shows estimated carbon storage in above ground vegetation—old growth forests, in particular, store large amounts of carbon.

Table 4. Overland Carbon Density of Various Land Uses[231]

Vegetation	Vegetation Tons CO_2/Hectacre[xxvii]
Urban and Suburban	0
Cropland	1
Mowed lawns/grassland	1
Rough marsh/swamp	2
Peat bogs	2
Shrubs/heath/moor	2
Young forest (1-10 years)	2.5-10
Old forest (100-plus years)	65-133

A clear way to compensate for our emissions, including the clear-cutting of forests, is to plant more trees and/or mangrove forests. However, to store additional carbon over the long term, we could plant trees that are fire resistant (fires will be more prevalent as the earth warms), and we should plant them in such a way that they can get old.

---

xxvi When a tree dies, it eventually decomposes, and this releases CO_2.
xxvii 1 hectacre = 100 acres

Therefore we could plant more trees that are fire-resistant (baobabs, larch, redwood, etc.), and plant trees that produce fruit or have other uses, so that local people will not cut them down[232]. Similarly, there is an "agro-forestry" movement[233] that advocates planting forests in plantations (where the wood is eventually cut, and regrown), or planting trees in combination with other crops[234]. However, it is an open question whether the world can increase the carbon density of our modern landscapes enough to compensate for Amazon deforestation, given the population pressures shown in Figure 23.

Carbon Capture and Storage Summary

Carbon capture and storage has not yet been done at large scale for coal plants, and preliminary cost estimates show $20-$70 a metric ton at large scale, which has not been successfully demonstrated[235]. When one adds in serious "not-in-my-backyard" issues, as people will not want to live near CO_2 storage sites, it is clear that carbon capture and storage is not an attractive or practical option. Similarly, biochar schemes suffer from higher $300 to $500 a metric ton costs. We could plant fire-resistant trees instead of excess corn on 90 million acres for emissions-increasing ethanol. But this would only sequester .6-1% of our annual emissions. Overall, carbon capture and storage schemes simply do not appear practical.

Strategy #4: Clean Up Afterward

As discussed above, and shown in Figure 38, direct air capture of CO_2 appears to be attractive, except that it is currently expensive to separate CO_2 once it is mixed into the atmosphere at lower concentration. In the words of Socolow: "We should be suspicious of distractions — direct air capture is one of these... At the top of the agenda should be decarbonizing the global power system. There is something grotesque about pulling CO_2 out of the air in one place while putting it into the air at 400 times greater concentration at another place. First things first." [236]

However, if we have already arrived at a tipping point, and Earth's natural CO_2 removal processes are saturated, we may have to use some method of direct air capture to remove CO_2 (or methane) from the atmosphere. In direct opposition to Socolow's views above, direct air capture does have a transportation advantage because a direct air capture machine can be operated anywhere.

A simple CO_2 direct air capture method is to pump air through sodium hydroxide (NaOH), a common industrial base chemical, which creates liquid sodium bicarbonate. The problem is how to economically recycle the liquid sodium bicarbonate back into sodium hydroxide. This has not yet been solved, so current "clean up afterward" methods cost from $100 to $1,000 a kg, which is equivalent to 3%-30% annual U.S. GDP.

Action Plan Summary

To summarize, the Economic Law of Pollution and lean engineering advocate the following short- and long-term plans:

Short Term

1. Support higher CAFE standards to truly double MPG (through aerodynamics, light-weighting (which is not connected to safety), engine efficiency, all-electric, and hybrids, as decided by the market). Current CAFE standards suffer from test method inaccuracies and loopholes, and these should be closed. These standards are relatively inexpensive, as the fuel savings will pay for near-term added costs over the long term, and they mitigate energy cost myopia.

2. Per strategy #1 of the Economic Law of Pollution, improve energy efficiency to reduce total energy consumption, and therefore CO_2 emissions. "Nega"-

watts[xxviii] is one of the cheapest sources of energy, but also can be limited in its overall effectiveness (i.e. it can easily reduce consumption 5-10%[237], but further reductions are asymptotically more difficult to achieve economically as the costs of efficiency improvements increase).

3. Substitute cost effective renewables for coal, up to 30%-40% penetration levels.

4. In areas of the country with higher coal usage that would exceed these penetration levels, promote switching of coal to natural gas, as this is highly cost effective—but not over the long term[xxix].

> Utilities continue to switch to natural gas, as it is cheaper than coal despite higher initial capital costs. The government could potentially accelerate this transition by offering natural gas capital equipment subsidies (low interest loans, etc.)[xxx] — being careful to promote this only *after* cost effective renewables are in place.

> As the EPA is doing, promulgate regulations to ensure that stray methane emissions are captured (measure these emissions very accurately, including stray and intermittent emissions, throughout the value chain and lifecycle of natural gas usage, and add regulations if (and only if) necessary, recognizing that the industry already has financial incentives to achieve zero emissions (emissions equal revenue loss).

[xxviii] "Nega"-watts is the use of energy efficiency to reduce demand instead of building a new power plants.

[xxix] While natural gas produces half the CO_2 of a coal plant, it still produces CO_2. Simple substitution will not get to greater than 50% reduction levels, especially in the long term.

[xxx] Taking a careful look at upcoming coal plant retirements, and whether such subsidies would actually have an impact.

5. Finally, changing farmer practices (e.g. improving tillage, reducing/improving fertilizer usage, planting more trees alongside crops) can reduce U.S. emissions by 3% at reasonable cost.

However, the combination of these 5 ideas will only get us to about 35% reduction, as shown in Figure 42. It is a good start, but will not be enough to support economic growth as our fossil fuel stocks run out over the next two generations.

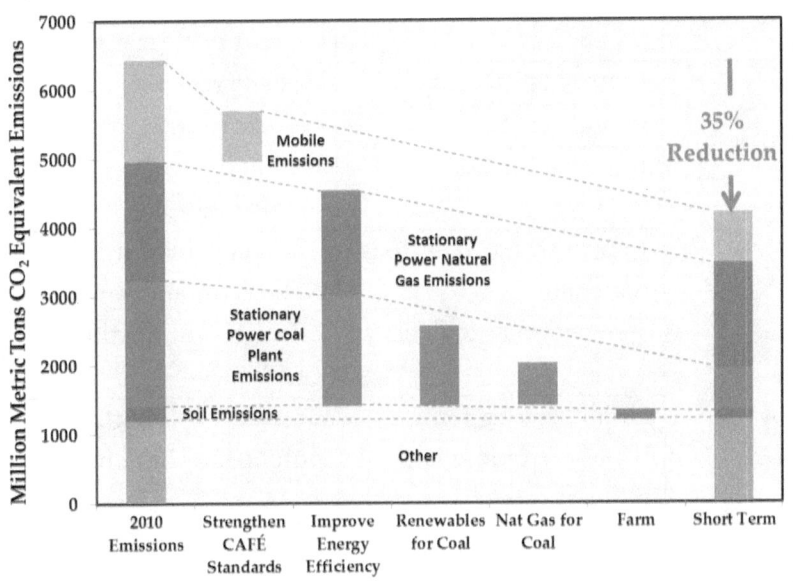

Figure 42. Short Term Action Effectiveness

Long Term

To reduce the cost of decarbonizing our economy, we need to do the short term actions above to buy us time, combined with R&D to tackle the remaining technical and economic barriers. When we succeed in addressing these issues, non-fossil fuel sources will be competitive, allowing natural substitution of these sources by an activated market.

Table 5. Long Term R&D Decarbonized Economy Transition Tasks

	Task	Purpose
Technical	Higher energy density energy storage	To enable gasoline equivalent range for electric vehicles
	Reduce energy storage cost	To enable variable renewable power source dispatch-ability[xxxi]
	Breeder nuclear reactors	More effectively deal with current nuclear waste
Cost	Improve efficiency	Reduce renewables electricity cost, to allow for energy storage, and eliminated subsidies
	Reduce cost	
	Improve lifetimes	
	Improved bio-feedstock mass per acre/volumetric density	To enable bio-based fuels to be cost- and land-use effective, to substitute for gasoline
	Establish long term benefits of bio-char, reduce costs	To fully explore the potential of bio-char at scale
	Improve efficiency of water electrolysis/solar production of hydrogen	To enable economical production of hydrocarbons
	Improve direct air capture of CO_2, efficiency and cost	Tipping point insurance to reduce CO_2, albeit expensively

---

[xxxi] Dispatch-ability allows a utility to give a customer power when they want it rather than when it is available. For example, solar power produces power during the day, and not the night. For power to be truly dispatch-able, we must be able to store it until it is needed.

The goals for this research are to reduce "at scale" costs to be competitive with fossil fuels, which allows eventual substitution for fossil fuel and continued economic growth.

Comparison to Alternative Plans

A number of "super decarbonization" plans exist. Amory Lovin's book "Reinventing Fire[238]" applies the Economic Law of Pollution to reduce waste and improve efficiency in the transportation, building, industry, and electricity generation sectors. He emphasizes that efficiency improvements come first, because they are least expensive. Subdividing by Economic Law of Pollution categories, his ideas include:

Economic Law of Pollution Category	Transportation	Buildings	Industry	Electricity Power Generation
Efficiency/ Utilization	Ultra-light auto bodies; improve auto aerodynamics; low rolling resistance tires; blended wing bodies to improve aerodynamics	Smart windows; radical insulation; enhanced evaporative cooling; phase change materials; LEDs; integrative design	Reduce process energy usage and distribution losses by improving efficiency of energy conversion devices	
Substitution	Electric propulsion; simplifying drivetrains; improve battery storage; hydrogen or natural gas vehicles; aviation biofuels		Combined heat and power; natural gas for coal; solar or wind for natural gas	Nuclear plus coal CCS; renewables; and more distributed renewables
Capture While Emitting				
Clean Up Afterward			Recycling	

Key barriers to the hyper-efficiency plans in "Reinventing Fire" include those already touched on: energy cost myopia, low energy prices, renter/owner "who gets the benefit" issues, and legal/utility barriers to combined heat and power (CHP) and renewables.

The U.S. Deep Decarbonization report[239] goes further, advocating electrification of most end uses (including space heating by heat pumps and water heating through electric hot water heaters). It recommends using electric and/or hydrogen vehicles (with H_2 produced by electrolysis of water or natural gas reformation with CCS) for transportation. For energy storage, it assumes electric vehicle charging, thermal building loads, pumped hydro, and natural gas with CCS. Costs are estimated at 1% of U.S. GDP, with huge uncertainty.

A study from NOAA[240] shows that wind and solar penetrations can be increased to 40%-55% if the U.S. were to build an optimized large scale HVDC transmission system backed by nuclear, hydro, and natural gas.

David MacKay, in his book "Sustainable Energy—Without the Hot Air,"[241] proposes a number of similar decarbonization plans for the UK, which has a population density much higher than the U.S., and a much lower land area. Electrification of transportation improves energy efficiency by 4 times, but requires doubling or tripling electricity generation capacity. Pumped storage and/or links to Norway to share their pumped storage are two other ideas to improve energy storage.

Similarly, Stanford Professor Mark Jacobson and colleagues have produced plans for each U.S. state to be 100% renewable[242]. Again, everything is electrified, including transportation, heating/cooling, buildings, and industry. Because thermal losses in power plants and cars are eliminated, total energy use goes down, producing a savings[243]. Battery-powered light duty vehicles are assumed, and hydrogen fuel cells are assumed for heavy duty vehicles and long distance shipping, and electrolytic cryogenic hydrogen is proposed to fuel aircraft[244].

Table 6. 100% Decarbonization Technology Status[245]

	U.S. Cost (vs. fossil fuels) w/o subsidy	Energy Density			Energy Return on Energy Invested (EROEI)	Lifetime (years)
		MJ/ Liter	MJ/ kg	Acres/ Gwhr/ year		
Onshore Wind	Competitive			20-40	18-25	10?-20?
Photovoltaics	On the cusp			6	~7-10	25
CSP Trough	Too $$$			3.5	21	30
Hydro	Lowest			25	100	50+
Geothermal	Med			1	3-5	20-30+
Nuclear	High			.5	5-15	40+
Coal	Med	6-22	10-24	1.5	78-82	30+
Natural Gas	Low	.036	55	1.5	8-12	30+
Oil/ Gasoline	N/A	32	44	N/A	15-20 (imported)	~1-3
Lead acid Battery	Too $$$	.18-.36	.04-.14			3-10
Li-ion Battery	Too $$$	1.1-2.4	.4-.9			~2-3
CAES	High	.007-.02				15-40+
Pumped Storage	High	.003-.007				50+
Flywheel	Too $$$	.07-.29	.2-3			20+

However, these plans are not realistic *at this time*, due to three factors: Non-fossil fuel energy production and energy storage concentration is too low, economical and effective energy storage does not exist (especially for vehicles), and the EROEI[xxxii] of renewable sources is too

---

xxxii See Appendix F.

low when combined with the energy storage required to compensate for their variable power output.

Because of the above issues, "Beyond Fossil Fuels" author Leo Smith advocates that nuclear is the only answer for the UK; Robert Bryce in "Smaller Faster Lighter Denser Cheaper" also argues that nuclear is the only answer, as does James Hansen[246]. They suggest that nuclear costs are reasonable and tractable, despite Fukushima Daiichi, Chernobyl, and Three Mile Island, and that nuclear power is the only viable way to lower carbon emissions quickly. For transportation, Smith argues synthetic fuels and/or electric vehicles will be needed; Bryce suggests natural gas, including vehicles, can be included in the mix.

The nuclear industry's safety record is good relative to other industries, as all three major accidents above were preventable[247] and the total release of radiation has been low overall. But despite our best human efforts, these releases did occur, in the most technically sophisticated nations in the world, due to human error. We are fortunate that the ocean is so big that it has diluted the Fukushima radiation to safe levels[248].

However, because of these accidents, the costs of nuclear energy —in the form of higher and more expensive safety standards —continues to climb. Nuclear plant construction costs also continue to climb because of high levels of fossil fuels used in construction (similar to roads), and the custom nature of nuclear plant designs[249]. Finally, decommissioning and storage costs continue to be infinite/undefined, as no solution to the long term storage problem has been found.

In the meantime, it makes sense to bury our nuclear waste thousands of meters below ground with all of the deep horizontal and vertical holes improved drilling techniques. These areas would then be below all water tables, in hidden locations that terrorists can't locate, and would be much safer than the large number of temporary few-meter-deep storage depots we have now, as many are close to rivers used for plant cooling water. Expansion of the Waste Isolation Plant Project, with salt deposits being safer because salt will fill in or seal any leaks, also makes sense. But even this project suspended operations in 2014 due to a human error radiation release, and isn't scheduled to come back online until 2016[250].

Ultimately, the long term nuclear vs. renewables vs. fossil fuels question should be decided by the market through unfettered market competition between energy sources. Eliminate all subsidies for renewables (i.e. the ITC); for fossil fuels (i.e. exploration and research subsidies); and for nuclear (i.e. repeal the Price Anderson Nuclear Industries Indemnity Act), and let the best technology(s) win. The market has shown itself to be highly capable of distributing resources cost effectively and efficiently.

V

How to Pay

"When learning is purposeful, creativity blossoms. When creativity blossoms, thinking emanates. When thinking emanates, knowledge is fully lit. When knowledge is lit, economy flourishes."
— A.P.J. Abdul Kalam, Indomitable Spirit

In America, a key barrier to fossil fuel reduction is economic—it will cost money to find and deploy fossil fuel alternatives and solve global warming. With the recent financial collapse and recovery, and with America's middle class becoming poorer because the price of fossil fuels and electricity is increasing, where can the money come from? Where is the invisible waste of cash, so that we can measure and reduce it?

Economics Background

As a prelude to finding the invisible waste in our economy, let us examine our wealth sources and how our economy grows. What are economic best practices, the drivers of growth and wealth creation, and the factors that can slow growth and destroy wealth?

These questions are large, and cannot be covered here in depth. Therefore, I summarize below my own understanding, with references if the reader desires to go deeper.

Economic Growth Factors

Economic growth models[251] posit that economic growth comes from the following sources:

(1) **Capital,** and how new investments in improvements, credit, and the presence of savings can accelerate growth.

(2) **Labor,** where increases in population and/or productivity lead to increases in market size and economic capability.

(3) **Technological progress,** or productivity growth, which includes any permanent improvement in the efficiency of production. As discussed previously, lean engineering efficiency, scale, and paradigm shift innovations drive our productivity growth, including creative destruction when technologies become obsolete.

(4) **Energy:** Because the level of wealth and culture in a society is determined by the net energy available per person[252], a corollary of White's law is that economic growth requires expanding the net energy available per person[xxxiii]. This means, fundamentally, that energy is getting cheaper when an economy is growing.

Note, #4 is under-represented in classic economic theory, as we Americans take cheaper energy for granted.

Economic Brake Factors

The converse of these sources of economic growth also applies.

(1) Negative Capital: While debt can be beneficial in the short term, excess debt — such as being incurred now by the U.S.

[xxxiii] With higher efficiency of production, technological progress can free up resources that then expand net energy per person. For example, if a more efficient car uses less gasoline, the gasoline not used will be used by someone else (perhaps in another country), increasing net energy available per person overall (worldwide). So points (3) and (4) can overlap somewhat, with respect to energy.

government —can slow our economy because of extra taxes needed to pay off the debt. Financial crises have a long history of causing recessions, or negative growth[253]. These crises have included overexpansion/debt defaults, crises of confidence and credit contractions, debasement/tightening of currency, inflation/deflation, speculation bubbles (land, commodities, stock, railroads, etc.), and trade wars and restrictions.

(2) Labor: Population decline can reduce economic growth as market size shrinks, which Japan and Germany have recently experienced and China soon will. Technology productivity can compensate for population declines.

Reduced labor productivity in a capitalistic system can slow economic growth in a number of ways:

a) Reductions in competition, including trade barriers, monopolies, higher market barriers to entry, land-use restrictions, labor market inflexibility, etc.
b) Inefficient regulations, complexity, and waste, relative to costs and benefits.
c) Market externalities can lead to reduced growth. One example is China's water and land pollution that reduces quality arable land in China; this increases local food costs, thereby reducing growth somewhat.
d) Concentrations of power/wealth and corruption. Corruption is theft, which directly reduces economic growth. High income inequality leads to low levels of savings because the poor save a very small proportion of their income, which reduces growth. When both are present, this can lead to high-income capital flight, which reduces capital available for growth.

e) Lower levels of education and lower quality of education reduce labor availability and quality, thereby reducing growth.

(3) Productivity Growth: Denard scaling,[254] which governs ever-faster speed increases on computer chips, stopped increasing in 2005. Moore's law, and the innovation it sparked, will most likely end by 2022. "It is important to realize just how odd semiconductor scaling has been compared to everything else in human history. ... From huts to skyscrapers, we've never built a structure that's thousands of times smaller, thousands of times faster, and thousands of times more efficient, at the same time, within a handful of decades." [255]

While Moore's law, and the computer revolution that it sparked, may be coming to an end, our modern society and lean engineering continue to improve productivity—the Internet gives more power to the individual, energy efficiency throughout the economy continues to improve, trade and competition optimize efficiency via Adam Smith's invisible hand of the market, and we continue to apply lean business practices throughout our service economy.

With regard to technical innovation and productivity, growth barriers can include:

a) The need for effective intellectual property protection to incentivize innovators[256]. It will be interesting to see how China's disrespect of intellectual property rights will influence its R&D innovation rate over the coming century.

b) Insufficient competition among projects/ideas. Modern R&D programs use "stage gates" to winnow product ideas down as they move from the conceptual stage to prototyping to final production. At each stage, the more ideas that are considered and discarded, the higher the quality of the products that get to final production[257].

Cluster Economics—Infrastructure for Growth

Michael Porter popularized the notion that sustainable competitive advantages accrue to businesses in clusters —whether geographical (the Pacific Northwest and aerospace), sectoral (automobiles in Detroit), or vertical (steel manufacturers near Detroit). Other examples of successful clusters include Hollywood (movies), and Silicon Valley (entrepreneurial computer technology companies).

Over the past few decades, cluster economists have examined supply and demand throughout industry value chains to find out which factors are critical for cluster and industry growth. With some overlap with the above factors, cluster economics teaches that there are seven foundations of growth[258]:

1)	**Labor skill**	The labor force must have education, industry skills, job training, and experience.
2)	**Innovation**	R&D know-how, and the presence of innovators.
3)	**Finance**	Sources of financing to accelerate growth, including a healthy banking and investment system.
4)	**Logistics**	Raw materials and finished goods must be able to get to and from world markets. Required infrastructure includes roads, ports, trains, etc.
5)	**Resources**	Regional resources can lead to growth; for instance, Norway's fjords and hydroelectric plants led to aluminum industry growth there. Required utility infrastructure

		resources include electricity, fuels, and water.
6)	**Governance**	Regulations can spur or hinder markets and/or competition (see above). Infrastructures to enable capitalist markets include political stability, legal systems (property rights, contracts, effective courts, etc.), public services (police, fire, education, sanitation), and effective tax structures (efficient, not corrupt)[259].
7)	**Quality of Life**	Areas with high quality of life tend to be more attractive to workers, especially knowledge workers. Infrastructure concerns include health care, communication services (mail, radio, TV, Internet, cell-phones), cultural opportunities, and beautiful natural environments.

These seven factors can also be thought of as sustainable competitive advantages, economic inputs, foundational infrastructure for growth, and the reason why companies locate where they do. In addition, cluster economics teaches that it is easier to grow where a critical mass of an industry cluster already exists.

Globalization, and Free Trade

Free trade is a policy in international markets under which governments do not restrict imports or exports, as embodied by the EU and the North American Free Trade Agreement. Reduced trade barriers have globalized the worldwide market, allowing capitalist markets to improve more people's prosperity than ever before. Even though it does create winners and losers, free trade overall improves economic welfare[260]. Even so, most governments in the World Trade Organization (WTO), including the United States, still impose some

123

protectionist policies to support local employment, such as import tariffs/quota/taxes, export subsidies, or limit export of natural resources.

Finding Invisible Economic Waste

Given this background, to apply lean engineering to non-technical realms and be able to afford to shift away from fossil fuels, first we need to examine what we spend money on in the U.S., so that we can find the invisible monetary waste.

Entrenched political groups will resist change fiercely, so politically some of the ideas below will be difficult to implement. Nevertheless, the first step toward reducing our wasteful spending is to identify where the most potential is—to make the invisible visible. I therefore use a data-driven approach and lean engineering principles to reframe how we think about some sectors of our economy, and to start a conversation. Fighting waste worldwide is a clear path toward global economic prosperity[261], and below are the areas where the data indicates we could do better.

Just recently, the federal government started breaking down national U.S. GDP into categories, so that one can look at 80:20 Pareto[xxxiv] trends. However, there is only data for a few years, which is too little to be of use for gauging trends. Instead, let us examine the Gross Output for the U.S., which includes final sales as the GDP does, but also includes intermediate product sales; because of this it is sometimes termed Gross Duplicated Output[262]. Data is available from 1997-2012.

[xxxiv] The Pareto principle is a rule of thumb: 80% of the problem is due to 20% of the causes.

Table 7. U.S. Gross Output, Relative Growth, 1997-2012[263]

Top 10 Economic Activity Category	1997-2012 Relative Growth Rate
Resource extraction, i.e. energy & mining	400%
Computer revolution	300%
Management of companies and enterprises	275%
Educational services	275%
Social assistance	250%
Performing arts, spectator sports, recreation	250%
Water transportation	250%
Other transportation and support activities	250%
Health care	200%
Miscellaneous professional, scientific, and technical services	200%

An indicator of waste is growth beyond what we might otherwise expect due to inflation or other factors. Therefore, Table 7, showing relative growth, gives us a place to start looking.

As we can see, resource extraction (mining of coal, oil, and gas[264]) is the largest relative growth category (with the caveat that oil and gas prices decreased in 2013-2015). The computer revolution, sports, arts (more leisure time), other transportation, water transportation, and miscellaneous technical services growth show how our society is becoming more technically oriented, how our machines create more leisure time, and how global trade is expanding. However, from a relative growth perspective, education, company management, and health care all appear to offer opportunities for waste reduction in addition to resource extraction/energy.

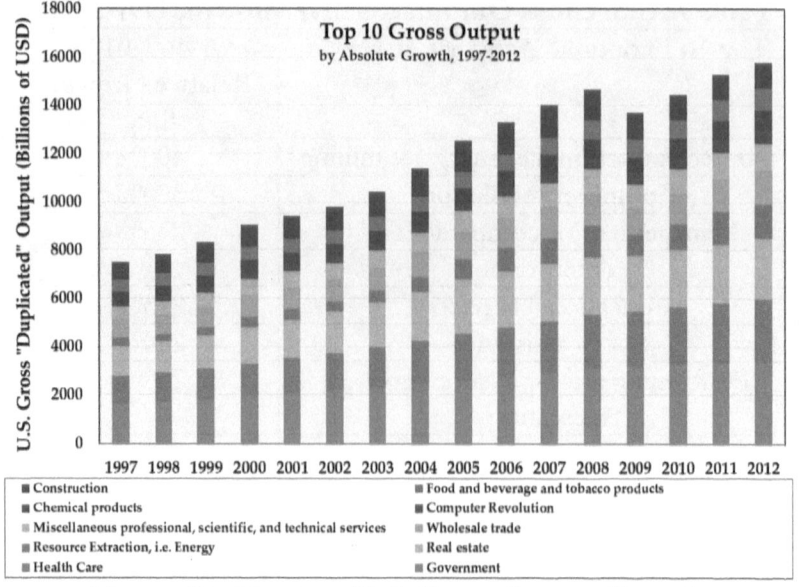

Figure 43. Absolute U.S. Gross Output Growth, Top 10, 1997-2012[265]

Figure 43 shows the same information as Table 7, but shows the absolute growth in Gross Output, rather than relative growth on a percentage basis. To the above three areas of waste, we can add one more, government. Let us consider waste reduction for these four areas, starting with the largest first. Note, as well, that each of these topics is worthy of volumes, so I will summarize my assessment and research into the root causes of the waste, in "lean engineering" fashion. For each waste category, I identify some actions that America can take to reduce the waste, with a goal of reducing costs by the 1% of GDP that it will take to start transitioning away from fossil fuels and solve climate change.

Government Waste

Our disaffection for politicians and distrust of government efficiency is at an all-time high, and continues to grow[266]. This is one key symptom of U.S. government waste, fraud, and abuse, all of which are jumbled in our minds as "corruption." But there is no common definition for these terms—which makes them hard to address. Fraud is defined by the U.S. government per Table 8. However, these accounting/financial definitions are the operating definitions for the Office of the Inspector General. They do not take into account some of the above lean engineering definitions of muda, muri, and mura, especially in regard to "Legislative" waste (conflicting objectives, overlapping agencies, start/stop workflow due to political funding fights, etc.) as opposed to "Executive" waste.

Table 8. U.S. Government Definitions of Fraud, Abuse, and Waste[267]

Fraud	Abuse	Waste
A type of illegal act involving the obtaining of something of value through willful misrepresentation.	Behavior that is deficient or improper when compared with behavior that a prudent person would consider reasonable and necessary business practice given the facts and circumstances. Abuse also includes misuse of authority or position for personal financial interests or those of an immediate or close family member or business associate. Abuse does not necessarily involve fraud, violation of laws, regulations, or provisions of a contract or grant agreement.	Waste involves taxpayers not receiving reasonable value for money in connection with any government funded activities due to an inappropriate act or omission by players with control over or access to government resources (e.g., executive, judicial or legislative branch employees, grantees or other recipients). Importantly, waste goes beyond fraud and abuse and most waste does not involve a violation of law. Rather, waste relates primarily to mismanagement, inappropriate actions and inadequate oversight.

Corruption can be defined as "the use of public office for private gain, i.e. use of official position, rank or status by an office bearer for his own personal benefit. Examples include (a) bribery (b) extortion (c) fraud (d) embezzlement (e) nepotism (f) cronyism (g) appropriation of public assets and property for private use and (h) influence peddling."[268] All of these somewhat overlapping definitions of corruption, fraud, abuse, and waste involve "waste."

From a lean engineering perspective, the goal is "added value"—any actions that do not add value are waste. But our government has many constituents it needs to satisfy, each with a particular definition of added value and mission, and each with goals for government and its functions. One person's waste is another's gold, a subjective value judgement.

This is a primary reason why lean engineering has been difficult to apply to government. That application is not a new idea. Newt Gingrich promoted Six Sigma to reform government as part of his campaign in 2012. How do you value public defense, or public education? How do you hold someone accountable to a subjective objective? How do you hold someone accountable when changes in top leadership occur every four years? In 2011, the American Society for Quality identified two other "lean applied to government" barriers: ongoing political partisanship, and the need to restructure the personnel management model used by many government agencies[269].

The Bible states that "For the love of money is the root of all evil: which while some coveted after, they have erred from the faith, and pierced themselves through with many sorrows"[270]. In 1887, Lord Acton said, "power tends to corrupt, and absolute power corrupts absolutely. Great men are almost always bad men"[271].

The age of these sources show that the world has been fighting corruption for a long time. To counteract corruption in modern times, Burmese economist U Myint proposes that we:

1) Reduce "economic rent," or monopoly profit—the extra profit that someone extorts from you by virtue of their advantageous position.
2) Reduce the number of laws, rules, regulations, and administrative orders that restrict business and economic activities, thereby generating economic rent. The rules should be transparent and consistent.

3) Reduce discretionary power. Because rules and regulations cannot be perfect, someone is granted discretionary powers to decide what happens. When this latitude is large, there is more room for corruption to occur.

4) Increase accountability. When those administering the rules are held accountable for their actions, corruption decreases.

5) Reduce market distortions of price controls, subsidies, and black markets. These distortions constitute a theft from one group to another (e.g. artificially low price-controlled rice means that rice farmers subsidize the city)[272].

Note that #2 and #3 conflict—more regulations are necessary to reduce discretionary power, but more regulations increase potential for economic rent. Less regulation (#2) increases discretionary power (the opposite of #3). Therefore a balance is needed. Also, low-income countries tend to have conditions that are more conducive for the growth of corruption[273], which in turn keeps them poor. The inverse corollary is that reducing corruption increases economic growth.

Despite these difficulties in applying lean engineering to government, the U.S. Department of Defense (DoD) applied it successfully to the Navy in 2006, then to the Army. In 2010, the Army produced $1 billion in cost savings and $3.3 billion in cost avoidances as a result of Lean Six Sigma (LSS)[274], and Forbes cited a combined $2.45 billion savings in 2008[275]. In 2013, the Army's business process and continuous improvement programs reduced costs by $2.6 billion, and the Army created mechanisms organization-wide to make continuous improvement part of "business as usual"[276].

So how can we tackle the largest waste in the U.S. government, applying the DoD's experience throughout federal and local government? Table 9 shows the current status of each of U Myint's corruption solutions as applied to the U.S. government, and some outstanding issues.

Table 9. Government Waste Status and Solutions

Government Waste Root Cause & Myint Solutions	Status	Outstanding Issues
1) Extra profit based on monopolistic position	• Anti-trust law prohibits monopolies • Regulate monopolies to reduce over-charging	• Some existing monopolies are not well regulated
2) Rules/law transparency	• Rules/laws are public • FOIA allows some viewing of federal decision making	• None • Freedom of Information Act (FOIA) is intrusive/distracting
3) Rules/law consistency	• Inconsistencies are decided by the judicial branch	• High level of delay in judicial system—inconsistencies exist for a long time
4) Number of laws/ rules/regulations	• 75000+ pages in the 2014 Federal Register[277] • Internal Revenue Code (IRC) is complex	• Regulations are onerous • IRC is onerous to comply with
5) Discretionary position	• 175000 pages in the 2013 Federal Register, limiting discretion • Separate judiciary decides, balancing executive branch	• Discretion is limited too much? • Judiciary delays create uncertainty
6) Accountability	• FOIA passed 48 years ago • Despite more public knowledge, no consequences for systemic acts	
7) Market Distortions	• Subsidies and special/exceptional treatment exists, distorting markets	• Subsidies continue to distort U.S. markets

Let us consider each of the solutions on this table.

Extra Profit Based on Monopolistic Position

America's founding was a revolt in response to the Tea Act of 1773 that enforced the East India Company's monopoly on tea. As a result, one of America's economic strengths has been awareness and vigilance with regard to reducing corruption and monopolistic power. America does a good job regulating natural monopolies that exist, and fostering competition via anti-trust laws where appropriate. The problem is when an invisible natural monopoly exists that is not recognized as such ... as was identified by a few of the waste areas identified by examining US Gross Output in Table 7 and Figure 43.

In particular, company management, health care, and education are all de-facto monopolies, as will be more fully explained below. We need to recognize this, and find better ways to regulate these monopolies because current methods of injecting competition into these spaces are completely ineffective.

Company Management

Executive pay compared to average worker wages has risen dramatically since 1983.

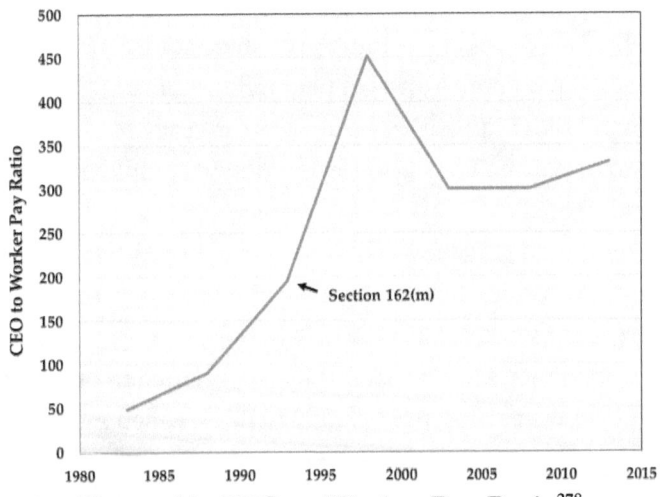

Figure 44. CEO to Worker Pay Ratio[278]

Proponents of high pay include three basic reasons why this level of pay is reasonable:

1) CEOs and higher level managers are responsible for leading large organizations, usually thousands of employees, and therefore should be highly paid.[279]
2) For shareholders, just a little bit better performance—say, 1%—can make a huge difference if billions of dollars are being managed. One percent of a billion is 10 million, so paying a lot for the best performance is worth it[280].
3) Technology in general has allowed corporations to grow much larger than in the past. By virtue of corporations' larger size and market capitalization, executives deserve more for this larger responsibility[281].

However, a number of studies have shown the opposite, that the recent rise in CEO pay is exorbitant:

a) Executive talent does not change or improve shareholder values, in general. Because executives are interchangeable, why should we pay more for one than another?
b) Executive pay has been decoupled from performance, even when only looking at long term shareholder profit. They have golden parachutes and get paid millions, even when they fail[282].
c) Shareholder profit is but one metric to measure CEO performance. They also have a duty to employees, customers, and products.[283]
d) There is an "Agency Problem"—the board sets executive pay, but executives have a lot of influence over their boards, so executives can set their own pay rates[284].
e) Studies have shown that raising pay relative to other CEOs will lead to ever-rising prevailing pay rates, called ratcheting[285].

It makes sense that a founder of a company like Apple should make multiple millions or billions of dollars. Rewarding entrepreneurs is a key driver of capitalism, and it is critical to keep this greed factor in place. But a CEO of a Fortune 500 company who simply inherited his company's structure, products, and earning power should only be paid 50 times the lowest paid workers wage, as was the case for decades prior to the 1980s. It is this difference between value creation and inherited wealth that should govern large executive payouts.

But value creation means different things to different people. Market capitalization is one measure, number and quality of jobs are others, improving revenue, profitability, return on invested capital, or cash flow are others. In principle, long term value creation should be more highly rewarded than short term.

If we limit prices artificially, such as with a maximum 12-times CEO to worker pay ratio that the Swiss recently rejected, it is the same as implementing price controls, which are well known to distort allocation of resources. Rent control reduces the quality and quantity of housing available; price ceilings also cause shortages (as seen in the 1970s when the U.S. imposed maximum gas prices); and price controls can be evaded by creating black markets, which increases corruption[286], as occurred during prohibition.

With regard to executive pay in the U.S, a 1993 law[287] specifically designed to limit executive compensation actually more than doubled executive pay in the form of stock-option-based compensation[288].

Therefore, we should:

1) Repeal this law, section 162(m) of the U.S. tax code, which has distorted the market through artificial price controls.
2) Create an independent rating agency that:
   a. Rates short term and long term private and public executive value creation performance based on:
      i. Stock investors (i.e. market capitalization)
      ii. Employees (i.e. sustainable quality jobs)

      iii.   Customers (i.e. maximizing customer value proposition, service, and revenue)

      iv.   Financial metrics (i.e. return on capital, profitability, etc.)

      v.   Environmental sustainability[289]

      vi.   Corporate social responsibility

b. Based on the ratings of thousands of executives, quantify ranges of performance in the above categories to define poor, steward/adequate, good, excellent, and amazing performance[290].

c. Evaluate and quantify all forms of executive compensation, forcing all companies to disclose CEO pay of all types to the independent rating agency, but not necessarily publicly.

d. Set maximum pay limits (including clawbacks/firing in the case of poor performance) for each category. Fifty times average worker pay for steward/adequate levels with maintenance value creation is appropriate, up to 1-5% of market capitalization for amazing performance. For large companies, such as GE, this limit may necessarily be lower.

e. However, these incentives should not be in the form of stock, as most investment advisors will say that it makes little sense to have your job income and your investment income coming from the same source—it is not diversified enough.

f. Have the government or another entity audit the third party rating agency to ensure objective evaluations are done.

g. Within the ratings ranges (i.e. X to Y million for good performance), have boards set incentives/compensation as is done now, and adapt the ratings to both company/industry specifics, and to executive specifics, including 360 degree feedback as appropriate. Boards should give feedback/input to the third party rating agency.

Right now, it is difficult for boards to compare executive performance apples-to-apples against others. For executives, better job performance ratings offer better, fairer, and clearer feedback. For shareholders, better executive performance ratings will reduce management costs and increase returns.

Such an independent rating entity would then also reduce vice-president and other executive pay—except in cases of high value creation—as these executives should not be paid more than the CEO, while still paying for amazing performance per a successful capitalist economy.

Health Care

The issue of health care in America is complex and out of scope for this book. Briefly, however, it is clear that the health care industry is a relative monopoly.

Hospital choice by consumers is relatively limited—typically one choses the closest hospital. Similarly, U.S. patients tend to stick with one doctor[291] because it takes time to build up enough trust for a good doctor-patient relationship. Medical schools—and congressional funding for medical schools and residencies—have limited the number of people graduating from medical school each year, despite a growing U.S. population. The reduction of graduates per million people since 1984 has limited the supply of doctors and specialists overall.

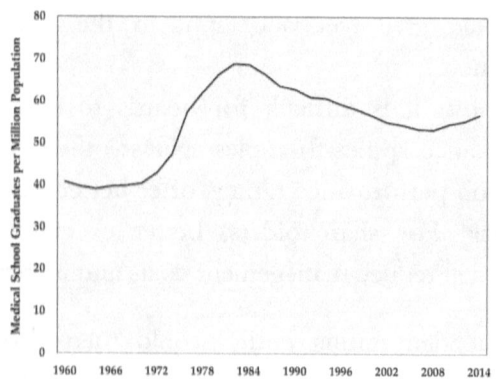

Figure 45. Medical School Graduates Over Time[292]

As a result, with limited competition and lower supply, physician salaries, and especially specialist salaries, are very high. All told, the U.S. spends five times more on specialist salaries than do other countries[293].

If we then consider higher pharmaceutical prices, high administrative costs for medical payments, and tort-based defensive medicine, as suggested by Todd Hixon[293], we end up spending almost 50%-300% more per capita on health care than do comparable countries, such as Japan, Germany, and Canada[294].

Other reasons for this excess include

1) There is a serious disconnect between health services and consumer lack of knowledge of pricing and options (when we have a health plan, cost is no object; when we don't, we use expensive emergency room care). We, as patients, don't know or care about costs.

2) We have higher obesity rates than the rest of the world, but different health habits in issues. For example, we have a lower number of smokers in the U.S.

3) Given physicians' monopolistic position and lobbying power, they are essentially setting prices and balancing the costs vs. benefits of all treatments. But managed care, where doctors are second-guessed by "bean counters," is problematic politically.

Individuals should be in charge of our own health, using doctors' advice, as they have the necessary knowledge and experience. But we have to recognize their (and the Association of American Medical Colleges) monopolistic power. This is the same as the case of public utilities, which are "natural monopolies" because the infrastructure required to produce and deliver electricity or water is very expensive to build and maintain. Similarly, the education of a doctor costs $380,000 for college[295] plus medical school[296], not including debt.

Monopolies are not new in our democratic and capitalistic society. We have anti-trust law to break up monopolies. And for true natural monopoly exceptions, Americans successfully regulate a number of industries, as shown by the utility industry, the insurance industry, and others. In general, this has taken the form of public utility commissions or insurance boards regulating prices to ensure profitable operations, reliable operations, and reasonable consumer prices.

But doctors will not willingly take a 50% pay cut … so what should we do?

1) Double or triple the number of open medical school slots to increase doctor supply, as well as eliminating the unsafe "doctor sleep deprivation is good" culture still prevalent in hospitals. Note that increasing the number of specialists will lower specialist salaries (via supply and demand), but not necessarily improve patient health[297]. Keep increasing the number of doctors until the U.S. reaches parity with the Organization for Economic Cooperation and Development on a doctor/million population basis.

2) In a similar vein, to increase supply and competition, allow more foreign doctors to be licensed to practice in the U.S. Over 30% of U.S. doctors come from foreign medical schools, but it requires years-long residency to practice in the U.S. Allowing nurse practitioners to increase the scope of their

services, as well as reducing telemedicine license limitations, would both help reduce the physician monopoly.

3) Following the example of utility and insurance regulated monopolies, create state-based commissions to regulate physician salaries, especially those of specialists. Ratchet down physician salaries, and reduce generalist-specialist differentials, gradually (say 5%-10% a year), until the U.S. joins the rest of the OECD. Pay generalists more, and specialists less[298]. In exchange for this regulation, fully publicly fund all U.S. medical schools and medical training beyond the current cap on Graduate Medical Education residency funding[299] (i.e. at roughly $700,000 a doctor times 18,000 doctors per year, this would cost about $13 billion, or .07% GDP. If we reduce doctor salary, the primary driver of health care costs, by 20-40%, the 20% of GDP we now spend on health care will reduce by far more than this .07% GDP cost.)

There is, of course, plenty more that needs to be done to fix Obamacare, and the health system in general, relative to tort reform, patchwork-quilt insurance paperwork and systems, efficiency, and pharmaceutical prices, but the above fix regarding physician salaries appears to be the first item on an 80:20 Pareto analysis of why U.S. health care costs are so high[300].

Education

On the education front, lean techniques of "Define, Measure, Analyze, Improve, and Control" have been applied to primary and secondary education in the U.S. through the No Child Left Behind (NCLB) Act of 2002. Standardized testing is used in schools, and corrective action applied when standards are not met. While standardized tests are not perfect—gifted student funding has suffered, more "teaching to the test" has occurred, and states have reduced standards—the law has brought a smidgeon of accountability to our educational system. It has also brought a lowering of standards, emphasized punishing failure rather than rewarding success, focused on absolute scores rather than recognizing growth and progress, created huge incentives to cheat, and fostered a "one-size-fits-all" inflexibility. Forty-one states have been granted waivers with regard to NCLB requirements. Lean engineering applied to education has simply not worked, and Congress amended it in 2015 to transfer more decision-making authority back to state and local governments.

If we compare the U.S. to other countries, the U.S. remains in the middle of the pack of the OECD, despite spending twice as much per student and having a relatively high level of educational inequity[301]. The NCLB standardized tests reveal where the poor and rich neighborhoods are, as schools are typically funded by local property taxes.

Schools have been constantly reforming (charter schools, merit pay, different standards, etc.) without actually changing much since the school system was instituted. We have the same teachers, in the same roles, with the same level of knowledge, teaching in the same schools, with the day organized in the same way, with the same courses and materials, with the same level of parental support[302]. We have seen many pilots that work, but fail to scale. School choice and vouchers have not made much of a difference, as is shown below in Figure 47. Teacher salaries have kept pace with inflation, but have not changed, as shown in Figure 46.

Insanity has been defined is doing the same thing over and over again and expecting different results. It has not helped to throw more money at education. Parents who pay for private education pay twice—once for public schools, and again for private tuition—but private school outcomes are not any better when adjusted for demographics[303]. To really change and improve our 1900s school system, it needs to be restructured.

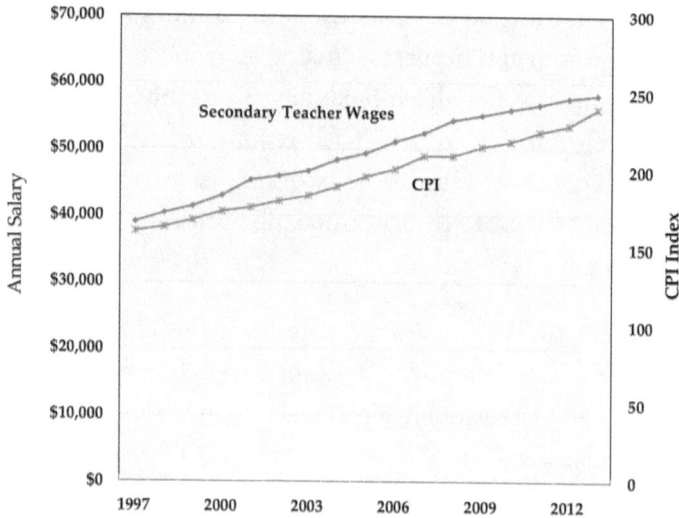

Figure 46. U.S. Secondary School Teacher Salary[304]

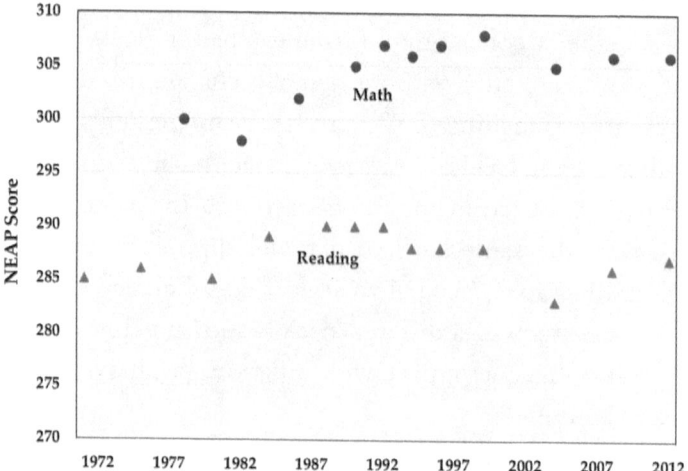

Figure 47. U.S. Assessment of Educational Progress Scores[305]

Or does it? When looking deeper, there are just as many myths and lies concerning American education as one finds regarding global warming.

Myth[306]	Truth
International tests show the U.S. educational system is falling behind	Educational achievement is strongly correlated with environment—cultural emphasis on education, family wealth, etc. When these are factored out, the U.S. system looks better.
School choice works, and charter schools are better	It really doesn't. Charter schools, with vouchers, sometimes look better because they get to screen applicants, and let public schools be stuck with less capable students.
Cyber-schools work	Actually, they are the worst type of schools, with low retention rates and poor performance—similar to correspondence courses of 50 years ago. "Cyber" is a useful adjunct for teachers, but a computer cannot motivate students to learn.
Teacher and school performance can be measured with standardized tests	Because educational achievement is most strongly correlated with environment (see above), measuring student performance is a highly imperfect measure of teacher or school performance. In addition, testing is highly variable year to year, depending on mood, individual motivation, environment, etc., and these are factors schools cannot control.
Unions are to blame	Only 35% of teachers belong to a union, which have limited power. States and countries with strong unions tend to slightly outperform their non-union brethren[307].

While it appears that while U.S. educational spending is on the high side, the international two-times spending difference per student is due to different wage levels, which vary linearly with total GDP (see Figure 48 below). A glance at any local school budget shows that administrative, building, equipment and other costs are dwarfed by teacher salary costs, as is appropriate.

And yet, the U.S. gross output statistics show that U.S. education spending has grown 40% from 1997-2012. So if teacher salaries have kept pace with inflation, where has our spending gone? And where and how can we improve?

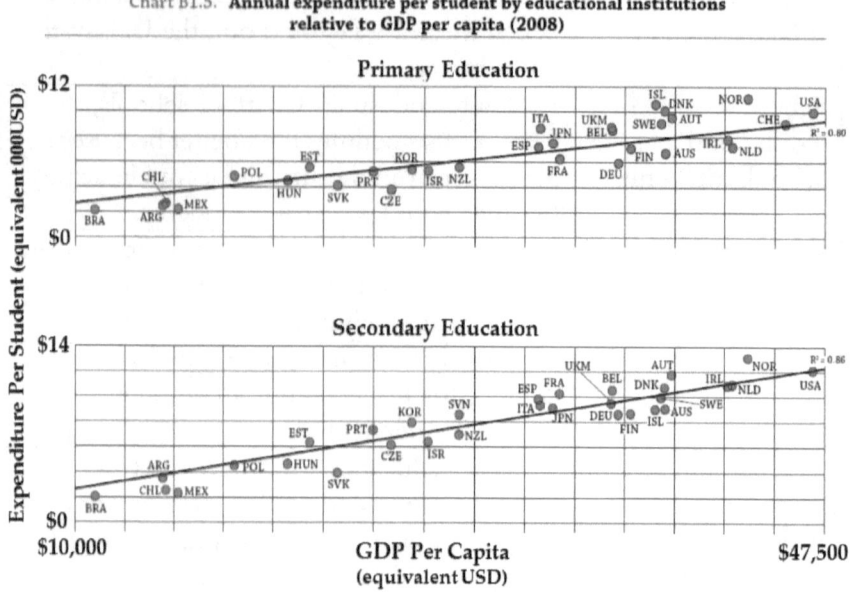

Chart B1.5. **Annual expenditure per student by educational institutions relative to GDP per capita (2008)**

Figure 48. Educational Spending per Pupil per capita[308]

When I talk to teachers, and reflect on my own experience as a parent of two, a few issues arise that lead to the overspending shown by the U.S. gross output statistics:

(a) Special education spending as a percentage of total education spending has risen from 4% in 1967 prior to the Individuals with Disabilities Educational Act (IDEA), to 17% in 1991[309], to 20% in 1999-2000[310], to likely something higher now[311]. Due to the fear of lawsuits, special education "has largely been insulated from considerations of cost and cost-effectiveness"[312].

(b) Like all government services, education is generally managed by an annual budget cycle. Educational professionals (teachers and principals) do not get rewarded for efficiency—if they

142

spend less one year, future budgets are likely to be reduced permanently.

(c) Teachers have a very hard job, and circumstances beyond their control (socioeconomic factors affecting their students, parental support, student motivation, etc.) have been shown to have a larger impact on student outcomes than they do[313]. This high level of difficulty, union seniority-based tenure, and a cooperative "all for one, one for all" attitude among teachers mean that those who perform poorly in any one year can continue to teach for a long time. So, while not common, there are teachers—maybe 5-10%—who have shown a pattern of poor performance and still continue to teach. Competition between teachers, or between schools, is stifled[314].

(d) As mentioned, private schools are doubly expensive because one pays for public school through property taxes, then pays for private school on top of that, but this affects only about 10% of the U.S. student population[315]. Expenditures per pupil also vary widely depending on geography, leading to wide disparities in teacher-student ratios, and in educational quality overall.

Solutions are relatively simple, but difficult to enact politically:

1) Fund all schools equally, no matter where they are located. This would require a change in how property taxes are used to pay for education.

2) Standardize how teachers get trained, attracted, prepared, supported, rewarded and retained. Top Programme for International Student Assessment (PISA) countries draw teachers from the top third in academics and give them extensive practice-based training.

3) Measure and value growth along other axes—social, emotional, ethical, and physical—in addition to cognitive.

4) Increase accountability and feedback to include professional judgment, and balance this with student testing.

5) Change the incentive system for teachers and administrators to reward efficiency, recognizing that what works in one place will not work in another. Base the top 10% of administrator and teacher salary on demonstrable continuous improvement in teaching cost effectiveness—with stretch goals that reward super performance.

6) The American with Disabilities Act has improved the lives of our special education and disabled population tremendously, with the power of potential lawsuits used to prompt societal change. But this prompt does not allow for consideration of cost effectiveness—we should study and model special education costs, and establish minimum standards for such things as aide to student ratios, student to teacher ratios, expenditures per student, and other key cost drivers. Such minimum standards could then define what is required by law, and save the country billions without compromising special education gains.

In summary, we need to better regulate our invisible monopolies, U Myint's first root cause of government corruption. Let us now examine the other root causes shown in Table 9 on page 130.

Rules/law Transparency

While not perfect, the U.S. has one of the most open and transparent government systems in the world[316]. Congress continues to deliberate Freedom of Information Act (FOIA) improvements, as shown by HR 653, S. 337 and others. PIRG and many other organizations continue to monitor state and federal government transparency[317].

Rules/Law Consistency

While the United States ranks 19 out of 102 countries for the quality of its legal system in a recent worldwide survey, its scores relative to delays remain a weakness[318]. Congress should study the cost of these delays compared to the benefits of better funding the system and/or staffing the judiciary (as Republicans have blocked Obama's nominees during his term[319]); recent immigrant system delays are unacceptable[320].

Number of Laws/Rules/Regulations

In the last 10 years, the U.S. Federal Register has published, on average, over 75,000 pages of new regulations (including rules, proposed rules, notices, corrections, and presidential documents)[321]. The Internal Revenue Code is complex and riddled with exceptions, taking average individuals an estimated 26 hours to comply with[322], and was over 73,000 pages in 2013[323]. The federal tax code is progressive, with the wealthiest taxpayers nominally paying the most taxes[324]. Federal progressive taxes are somewhat offset by overall higher regressive taxes at the state level, resulting in a patchwork of tax "fairness" or "unfairness," depending on who and where you are[325]. But regardless of how fair the tax code is, it is clear that the U.S. has overly complicated and voluminous regulations and tax codes that are inefficient and waste tremendous resources[326]. The total paperwork burden in the U.S. averages about 9 billion hours annually over the last decade[327].

There have been perennial discussions on the need for change but little action. Given the prevalence of special interest groups, and fierce political contention, annual across the board cuts in the number of pages and complexity are a first step, without regard for efficiency. Similar to the recent budget sequestration process, complaints about "how blunt and stupid this is" will result in more refined prioritization over time. Continue this annual—say 5% or 10%—cutting process until regulation/tax complexity reduction targets are met (as measured by hours to file, speed of compliance with regulations, cost vs. benefit, etc.).

Discretionary Position

Given the voluminous regulatory burden on U.S. citizens, does it make sense to allow government regulatory agencies more discretion in their decision making? If one does, this also opens up the possibility of more corruption[328], so this looks like a case of "if it isn't broken, don't fix it."

Accountability

The U.S. ranks relatively high with regard to government accountability[329], so again, we have a case of "if it isn't broken, don't fix it."

Market Distortion

In Chapter III, it is proposed to eliminate numerous fossil fuel, renewable, nuclear, and other energy subsidies that distort the market. Other wasteful subsidies that distort our economy include:

1) The National Flood Insurance program insures homes and allows rebuilding of homes within high risk flooding areas. Insurers have sensibly exited these markets, and the government has stepped in, which increases liability for taxpayers; the GAO reports that this program is on its "High Risk List" for needing future taxpayer bailouts. More reform is needed[330].

2) U.S. agricultural subsidies of $10 billion-$30 billion. The 1998 Direct Subsidy in the Farm Bill was repealed in 2012, but commodity and crop insurance subsidies still impact how farmers assess risk[331].

3) Tax subsidies. One reason the personal tax code is so difficult to reform is that Americans' hatred of taxes means that we love tax breaks, even though they are subsidies. Corporate tax subsidies are another large kettle of fish, with 26 of the Fortune

500 corporations (including Boeing, General Electric, and Verizon) paying no federal income tax between 2008 and 2012[332].

4) U.S. fishing subsidies, while hard to find, appear to be $713 million per year between 1996 and 2004, largely dominated by fuel subsidies and fisheries research[333].

Lean engineering is focused on waste reduction, and we can save money as a country by eliminating the above wasteful government subsidies, allowing the market to reflect real prices for everyone. This is summarized below in Figure 49.

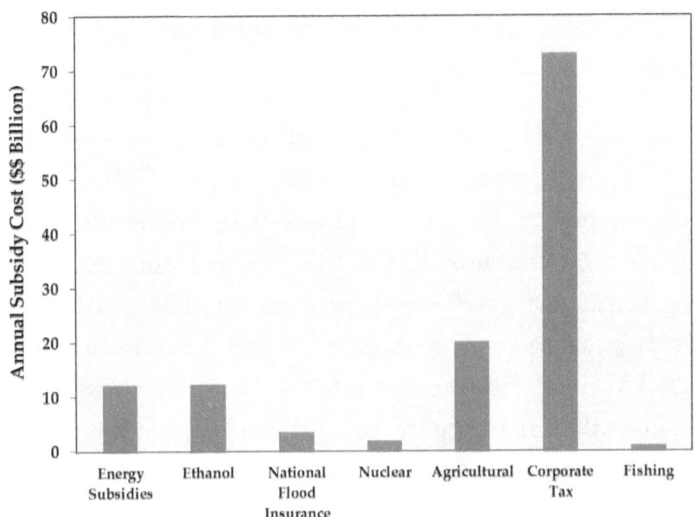

Figure 49. Subsidy Waste Estimates[334]

Promoting Economic Growth

In addition to finding and eliminating economic waste as outlined above, we can directly accelerate our economic growth. White's law, underappreciated in economic circles, states that economic growth requires expanding the net energy available per person. Lean engineering teaches us to reduce waste, and energy efficiency projects (such as insulating attics and windows, etc.) reduce our energy use by reducing energy losses. Such projects therefore increase the net energy available per person, and are invisibly responsible for at least 33% of our continuous economic growth rate since our country was founded. "A penny saved is a penny earned" (Benjamin Franklin). In the case of energy efficiency, a penny saved is a penny spent elsewhere to grow our economy.

Because energy efficiency increases our economic growth, we need to change how we evaluate the worth of energy efficiency projects. Currently, we use payback arguments in our heads, and our concept of "a penny now is worth much more than a penny in the future" means that we expect relatively low 2-3 year paybacks to induce us to spend funds on a project. In addition, based on payback-market penetration studies[335], only some of us think this way, as we are afflicted by "energy cost myopia."

But a better frame is to use energy efficiency *as much as possible* to increase economic growth—to the point where payback equals the economic lifetime of the product in question. Therefore, to promote this type of economic growth the U.S. government should give free financing for all energy efficiency projects that have payback periods less than or equal to average product lifetime (or equivalently projects with an NPV>0 @ 4% discount rate[336]). The interest expense by the government will be made up by the economic growth this promotes[337], and help mitigate energy cost myopia.

This myopia exists because we devalue future savings. If the government provides a way for energy consumers to monetize the future benefits of energy consumption in the present, this devaluation is eliminated. This is why direct rebates work better than zero or low-interest loans to promote energy efficiency. "Revenue neutral" fee-bates[xxxv], suggested by Amory Lovins in "Reinventing Fire," similarly eliminate our devaluation of future energy savings. To fully mitigate

---

[xxxv] A feebate scheme imposes a fine on high level polluters, and remits them to low level polluting consumers. For instance, for cars, one could impose a fee on high CO_2 emitting vehicles, and rebate this fee to buyers of new low emissions vehicles.

energy cost myopia, we should therefore allow equivalent direct rebates or revenue neutral fee-bates to substitute for "free financing."

"A Penny Saved is a Penny for Jobs"

Let's briefly show how this might work, with dummy numbers as I am not a government finance expert. The government creates a $10 billion fund that it either pays out direct 10% rebates, or spends this rebate on providing low interest financing, for monitored/verified energy efficiency projects that meet the NPV>0 @ 4% discount threshold[338]. For example, a factory is changing out motors on its production floor, but higher efficiency motors cost more; it proposes a project to replace its current motors with higher efficiency motors as they wear out. The government gives a 10% rebate for the motors, helping to economically justify the project, and the factory's electricity bill is reduced. Similarly, a homeowner, business, or schools may want to install more expensive insulation, LEDs, or other way to save energy. Many such projects, aggregated to a $10 billion fund, directly contributes to GDP at a 10:1 ratio ($10 billion spent by the government supports $100 billion in projects), which is taxed, obtaining some proportion back in taxes (2014 rates are about 12% sans Social Security[339]). These $100 billion in projects improve energy efficiency, saving at least 5% on U.S. energy bills. The money saved gets added to GDP, generating more taxes, resulting in a revenue neutral way to overcome energy cost myopia and promote economic growth. Consumers without ready funds can convert the rebate into a low interest loan.

Fixing "Split Interest" Market Distortions

Another market distortion in the energy efficiency space occurs when two entities' economic interests don't align[340]. A landlord provides a furnace, and the tenant pays the heating bills. If the landlord spends an extra $1,000 to get a higher efficiency furnace, the tenant gets the lower heating bills. So the landlord installs a low efficiency furnace to keep his costs down. Similarly, builders don't get the benefit of installing higher efficiency windows or insulation in the buildings they build, the building occupant does[341]. So cheaper windows and insulation are installed, locking in more wasteful energy use for the building's lifetime—even if the payback time might be less than 2-3 years.

These types of distortions exist because the market does not monetize the future benefits of energy consumption in the present. If the scheme above is followed, landlord/tenants and builder/occupant can split the direct rebate or revenue-neutral fee-bate.

Lean Engineering Savings Summary

In 80:20 Pareto fashion, by systematically, albeit cursorily, looking at the most wasteful portions of our economy, lean engineering techniques are applied to identify where America can save the most. This frees up funds to start moving away from expensive fossil fuels, mitigate global warming, and accelerate economic growth.

Examination of "gross output" for the U.S. shows that company management, health care, education, and government appear to offer opportunities for waste reduction, which is the primary focus of lean engineering. A synonym for government waste is corruption, and application of U Myint's solutions to corruption produced the following solutions:

Table 10. Lean Engineering Savings Action Summary

Company Management Monopoly	1) Repeal section 162(m) of the U.S. tax code, which had the unintended consequence of doubling CEO pay. 2) Create an independent entity that rates U.S. executives for value creation across multiple constituents, including stock investors, customers, and employees. Set maximum pay ranges for poor, steward/adequate, good, excellent and amazing performance, with current boards deciding actual pay within the range. The steward/adequate rating should be no larger than 50 times average worker pay, with amazing performance yielding 1%-5% of market capitalization.
Health care Monopoly	1) Double or triple the number of open medical school slots to increase doctor supply. Keep increasing the number of doctors until the U.S. reaches parity with the OECD on a doctor/million population basis. 2) Recognize that doctoring is a natural monopoly, and replicate the utility model into the health care space. Create state-based public doctor commissions to regulate physician salaries, especially those of specialists. Ratchet down physician salaries, and reduce generalist-specialist differentials gradually (5%-10% /year), until the U.S. joins the rest of the OECD. Pay generalists more, and specialists less. In exchange for this regulation, fully publicly fund all U.S. medical schools and medical training, which is the equivalent of high cost public utility infrastructure.

Education Monopoly	1) Fund all schools equally, no matter where they are located, changing how property taxes pay for education. 2) Standardize how teachers get trained, attracted, prepared, supported, rewarded and retained. 3) Measure and value growth along other axes— social, emotional, ethical, physical—in addition to cognitive. 4) Increase accountability and feedback to include professional judgment, and balance this with student testing. 5) Change the incentives system: Base the top 10% of administrator and teacher salary on demonstrable continuous improvement in school cost and/or teaching effectiveness— with stretch goals that reward super performance.
Education Monopoly (continued)	6) Create minimum standards for special education to rationalize costs while continuing to improve treatment of disabilities.
Government	Reduce the number of pages of regulations and the tax code by 5%-10% per year across the board until regulation/tax complexity reduction targets are met (as measured by hours to file, speed of compliance with regulations, etc.)
Market Distortion	Eliminate or reduce corporate tax, agricultural, ethanol, and fossil fuel subsidies.

Conservatively assuming 50% elimination of subsidies, and 10% improvement everywhere else, this gives a budget of approximately 1.4% GDP to spend on fossil fuel replacement long term research and climate change mitigation[342].

With regard to energy efficiency, the U.S. is already rated 13 in the world[343], so we do not start from zero. But, because of energy cost myopia and our focus on 2-3 year paybacks, there remains a lot of potential for energy savings[344]. "A penny saved is a penny for jobs" could therefore conservatively promote an additional .5% GDP economic growth.

VI

Timeframes: How to Accelerate Action

"Lost time is never found again."
— Benjamin Franklin

Climate change doubt and denial is on a collision course with scientists' warnings that global warming is real. Even corporations such as Exxon place a shadow price on carbon to estimate how much CO_2 pollution will cost[345], and Republicans in the Senate admitted at the beginning of 2015 that climate change is real[346]—although they deny that humans cause it[347]. As we bury our collective heads in the sand, we move closer and closer toward a "surprise" point where a catastrophe will make it clear that global warming is here, beyond the canaries in the coal mine enumerated previously. Denial may not be possible for much longer.

How Much Time Do We Have?
Whether or not such a surprise occurs, 90% of economically recoverable fossil fuels will be exhausted by 2070, as discussed in Appendix F. Beyond moral considerations, this chapter enumerates additional timing considerations that point us toward action now, as opposed to kicking the can down the road. These considerations include U.S. infrastructure turnover timeframes, the energy trap and technology development timeframes, fossil fuel stock "peaking" timeframes, climate change impact timeframes, laissez faire decarbonization, and wealth over time vs. cleanup cost.

Human Infrastructure Turnover Timeframes

A coal plant can last up to 40-50 years, so it takes that much time to change out our infrastructure. This can be significantly accelerated if we strand current assets[xxxvi], but doing so would increase costs by over 10 times. At $3,600/kW[348] for a coal plant and 308 GW[349] of coal capacity in the U.S., wholesale early replacement of U.S. coal plant capacity will cost $1.1 trillion, or 6.5% GDP. This is prohibitively expensive, so we need to get started now to slowly change our infrastructure.

The Energy Trap and Technology Development Time

Currently, non-fossil fuel sources cannot fully replace fossil fuels. They are too expensive, and we do not have energy-dense cost-effective storage technologies that can mitigate intermittency. Research to develop replacements will take some time, and it will take at least 5-10 years to move a technology from laboratory to full scale once a replacement is found.

In addition to being expensive on a dollar basis, current renewables are too expensive on an energy basis—their energy return on energy invested (EROEI[xxxvii]) is too low. Wind, with an EROEI of 20, comes closest, but this is still lower than a wind turbine's lifetime, so does not yield a fast net gain of energy to society. Solar, at 10-15, is similarly too low. For healthy economic growth, which depends on increasing net energy available to society, we continue to plunder and spend our fossil fuel stocks, and will do so for the foreseeable future. Competing with the "free" stored energy inherent in the Earth's compression and densification is difficult. Because it will take a lot of time to find solutions, we need to get started now.

[xxxvi] A stranded asset is one which is retired earlier than its nominal lifetime, and then written off.
[xxxvii] See Appendix F for further details.

Fossil Fuel Stock "Peaking" Time

As we have seen from every coal producing nation in existence, government estimates of "economically recoverable" coal reserves are always severely overestimated[350] until falling production forces them to say "oops." The U.S. is no exception, and academic studies predict that 90% of economically recoverable coal will be exhausted by 2070[35]. EIA estimates of how long oil and gas reserves will last at current usage rates also show the year 2070, a little over two generations[351]. By that time, the cost to extract fossil fuels will rise to the point where it is no longer cost-effective, or energy-surplus effective, to bother with further extraction. Because our modern economy is so tied to fossil fuel usage, our economy will collapse and severely contract per White's law.

It could happen sooner, or later—nobody really knows. While there is currently a glut of natural gas due to fracking, some skeptics and industry insiders believe that the shale "revolution" will only last 10-20 years[352]. But even if the shale skeptics are wrong, recent increases in extraction costs show that we still need to do something *now* to reverse this trend. These same actions to slow the use of fossil fuels and find viable substitutes will also extend fossil fuel industry profits, by boosting prices and volumes over the long term[353].

Climate Change Timeframes

Because the canaries in the coal mine are perishing, because the ice has started to melt during a natural cooling cycle, our fossil fuel CO_2 emissions have tipped the scale toward heating.

As discussed previously, because Antarctica and Greenland will likely take centuries to melt, we have at least eight generations, or about 200 years, to reverse these trends and save our sea-coast cities and Florida.

This long time span is comforting because surely we will figure out a solution to fossil fuels by then, if only by necessity because they will have run out. But we don't know when or if we will reach irrevocable tipping points in methane tundra release, ocean circulation change, etc., that might accelerate global warming. While less likely, we could also trigger unknown negative feedback loop mechanisms that will give us more time. Given the uncertainty, it makes sense to get started now on solutions.

Laissez Faire or "Copy Europe" Decarbonization

As a country, America has not focused at all on decarbonization of our economy, so it was somewhat of a surprise that we recently reduced emissions by 15% as fracking decreased the cost of natural gas power relative to coal, causing fuel switching in the electricity industry, and high gas prices reduced the amount we drive[354].

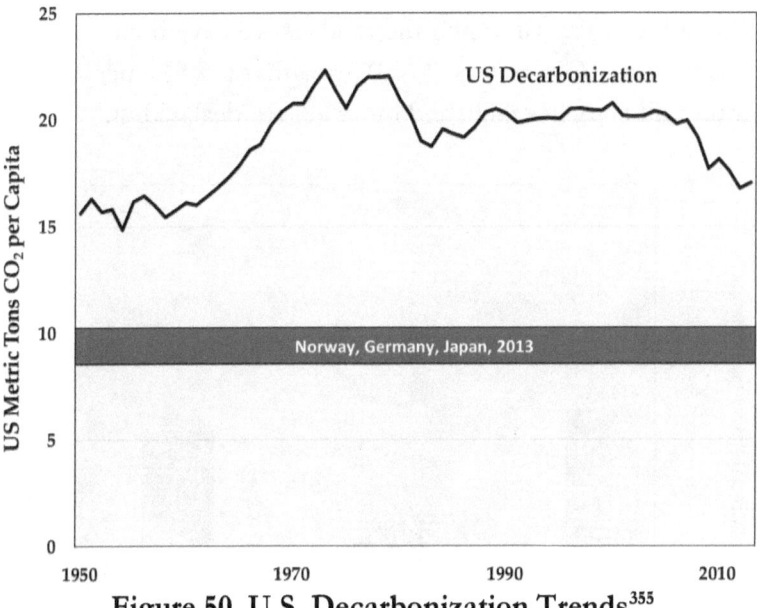

Figure 50. U.S. Decarbonization Trends[355]

With Germany, Japan, and Norway at annual CO_2 emissions between 9 and 10 tons per capita (per Figure 2), it is clear that we could, with some effort, reduce emissions per capita by a further 30%, as described in Figure 50. These three countries accomplished this decarbonization primarily through (1) higher electricity prices, which promotes conservation; (2) higher efficiency standards; and (3) use of nuclear power. With both Germany and Japan reducing their nuclear fleets sharply in response to Fukushima, their decarbonization trends will reverse, especially as Germany has turned toward using more lignite coal[356]. In the U.S., population is more spread out, with public transportation systems that are utilized less[357], so Americans will need to use more CO_2 for transportation. Nevertheless, from a time-frame perspective, Japan and Germany have accomplished their reduced emissions since their reconstruction after World War II, taking about 50 years. These countries have shown us how to decarbonize relatively painlessly.

Wealth Growth vs. Cleanup Cost

So why not kick the can down the road, as we have done for the last three decades? Given U.S. GDP growth of 2.5% per year, our economy will be richer then, and more able to deal with it.

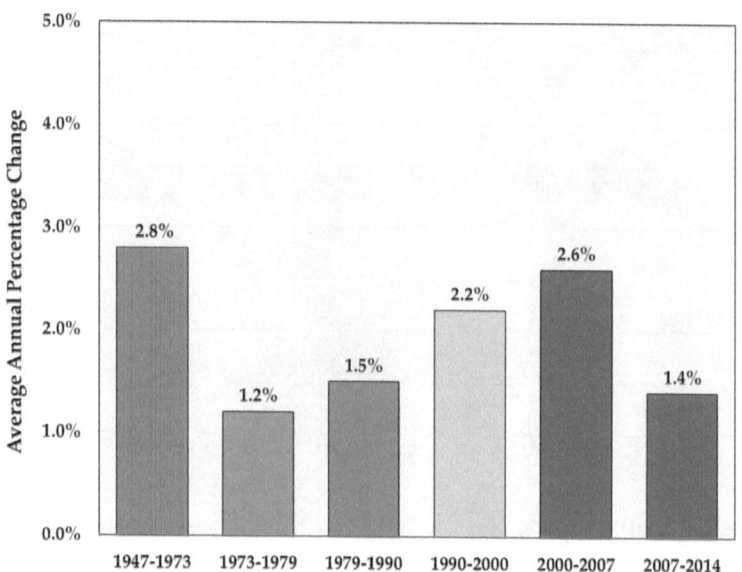

Figure 51. U.S. Productivity Change, Non-farm business[358]

Or will we? Some experts predict that economic growth will falter in the coming century[359], because U.S. educational attainment is projected to slow, population growth levels out, global warming pollution costs increase, Moore's law ends, etc. Labor productivity in the U.S. has dropped recently. Our service economy is harder than manufacturing to make more efficient, Uber not-withstanding.

What will replace Moore's law, now that the computer revolution has mostly rippled through our service-sector-based economy? Certainly globalization, to some extent, but the lifting of four-fifth of the world out of poverty will accelerate the world's CO_2 pollution, as the developing world copies our high energy using ways.

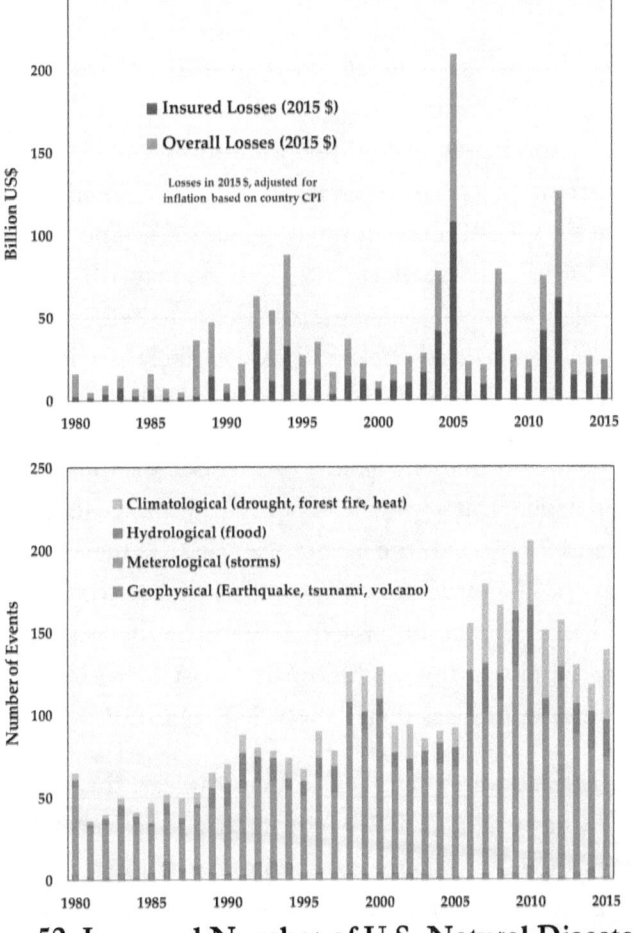

Figure 52. Loss and Number of U.S. Natural Disasters[360]

Then there are the costs of global warming, which, in general, will first affect the poor close to the coasts, similar to what occurred with Katrina in 2005, and Sandy in 2012. The insurance industry in the U.S. has recently changed its dismissive tune regarding global warming, as both losses and the number of natural disasters have been increasing.

Initially, adaption costs will be manageable. Modern agriculture can adapt to coming dislocations, sea level change will occur gradually, we will invent new ways to fight forest fires, and we will improve our storm preparedness and early warning systems. Nevertheless, these costs will mount. Estimates of adaption costs range from 1% to 4% of GDP[361].

But, eventually—whether in 50 years or 100 or 500—the costs of unchecked climate change will be more catastrophic. Bees will die off completely[362], forcing us to use miniature drones to pollinate flowers. The permafrost may emit large amounts of methane into the atmosphere[363]. Ocean currents will change[364]. The Amazon may decline or burn[365]. Extra heat and dryness may offset higher CO_2 stimulated plant growth, causing crop yields to fall[366]. Sea levels will rise[367]. While humans and cockroaches will likely survive due to high adaptability, life on Earth will become more unpleasant.

However, the best argument against further delay is the Economic Law of Pollution itself. Strategy 4, Clean Up Afterward, costs 10 times what Strategy 1 and 2 cost, and the longer we wait, the bigger the mess will be. Now we pollute at levels of 5.4 billion metric tons[368] of CO_2 per year (see Figure 31). As discussed in the next section, there may be methods to eliminate this pollution that cost $20-$25 a metric ton, which is equivalent to less than .8% U.S. 2014 GDP.

So, to simply tread water, we need to spend at least this much annually. But the U.S. has polluted 382 billion metric tons since 1800[369], a big mountain as shown in Figure 3. At $25 a metric ton, two years' worth of pollution (i.e. 10.8 billion metric tons) will cost 1.6% GDP annually and take 35 years to clean up. If we could double this rate, this would mean spending 4% of GDP, and taking 18 years. And, of course, if it costs $250 a metric ton, it will cost 10 times as much or take 10 times as long. Every year of delay will cost an extra .8%-8% of U.S. GDP in the end.

As we have seen, all environmental cleanup jobs[370] take years, decades, and centuries. The longer we delay, the more expensive it gets. CO_2 is no different, and the easiest way to climb a mountain is step by step, putting one foot in front of the other.

Timing Summary

The following diagram summarizes the four general timeframes discussed above that are relevant to global warming action: pollution production and reduction, technology delays, fossil-fuel peaks and costs, and global warming costs.

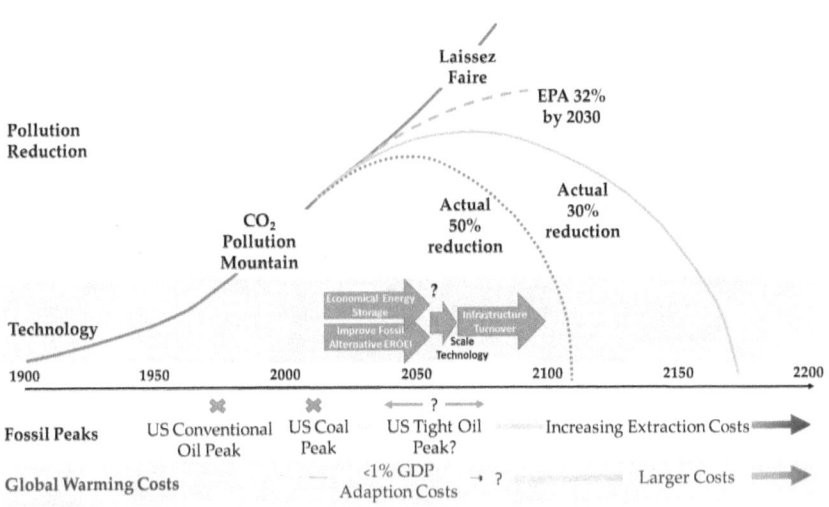

Figure 53. Global Warming Timeframes

It shows how our CO_2 mountain will keep growing if we do nothing to reduce our fossil fuel usage, and how global warming costs, fossil fuel peaks, and technology development times all come to a head in the next few generations. Given long infrastructure, scaling, and research timeframes, we are relatively behind in the necessary race to find fossil fuel alternatives.

How Not To Accelerate Action

Pacala and Socolow wrote a paper espousing the idea that "a portfolio of technologies now exists to meet the world's energy needs over the next 50 years and limit atmospheric CO_2 to a trajectory that avoids a doubling of the preindustrial concentration."[371]

Table 11. Climate Stabilization Wedges

		Feasibility/Comment
1	Doubling fuel efficiency of 2 billion cars from 30 to 60 mpg	Yes
2	Decreasing the number of car miles traveled by half	Overall vehicle miles traveled (VMT)/capita has reduced by 10% since 2005 but may trend upwards. Increasing public transportation may be difficult in the widespread U.S.
3	Using best efficiency practices in all residential/commercial buildings	Yes
4	Producing current coal-based electricity with twice today's efficiency	No, we're at a technical limit unless we employ CHP
5	Replacing 1,400 coal electric plants with natural gas-powered facilities	Yes, but not long term
6	Capturing and storing emissions from 800 coal electric plants	Too expensive, dangerous
7	Producing hydrogen from coal at six times today's rate and storing the captured CO_2	
8	Capturing carbon from 180 coal-to-synfuels plants and storing the CO_2	
9	Adding double the current global nuclear capacity to replace coal-based electricity	
10	Increasing wind electricity capacity by 50 times	Yes, but only up to 30%-40% penetration
11	Installing 700 times the current capacity of solar electricity	
12	Using 40,000 km^2 of solar panels (or 4 million windmills) to produce hydrogen for fuel cell cars	H_2 fuel cell cars too expensive
13	Increasing ethanol production 50 times by creating biomass plantations with area of 1/6 world cropland	No—leads to more emissions
14	Eliminating tropical deforestation and creating new plantations on non-forested lands to 5 times current plantation area	Improving the efficiency of our land use may solve deforestation issues. As agricultural efficiency rises, land is freed up to become forested again.
15	Adopting conservation tillage in all agricultural soils worldwide	Efficacy overestimated

The job of stabilizing emissions was chopped into pieces, each called a "climate stabilization wedge"; each wedge represented 1 billion metric tons of CO_2 emissions per year by 2054. As shown in the comment column, only 3 or 4 out of 15 wedges appear feasible. We should therefore focus on these, while doing research to find reasonable alternatives. But, while the concept of climate stabilization wedges is potentially useful to focus action, it encourages a scattershot approach that is ineffective, and dilutes the world's limited resources. Lean engineering instead encourages a focus on the largest problems first, applying the 80:20 Pareto rule.

For example, we could do a lot of scattered research to address the numerous but smaller components non top 3 emissions components shown in Figure 30 on page 83. As annotated in Appendix B, this research includes using electric arc furnaces instead of basic oxygen furnaces in steelmaking, cement alternatives, improved fertilizer delivery, and changing the microbiota in cows to reduce their methane emissions. But, in a world with limited R&D resources, guided by Pareto principles, these resources should be focused on the tasks shown in Table 5 on page 112..

How To Accelerate Action

In summary, to accelerate the transition from expensive fossil fuels, we must:

> Activate the market per Chapter III, with all technologies competing on a level playing field:
>> o Impose "EPA equivalent" import tariffs on products from countries that do not comply with EPA pollution standards, giving these funds directly back to the EPA equivalent in the foreign country. This will reduce our long term pollution cleanup costs and repatriate jobs.
>> o Evenly impose low level CO_2 pollution fines to reflect the cost of cleaning CO_2 from the air—but not high carbon taxes, which would lower economic growth.

- o Eliminate corn ethanol RFS mandates, fossil fuel exploration, depletion and cleanup subsidies, U.S. financing of overseas fossil fuel projects, Clean Air Act grandfathering of coal plants without emissions controls, the Price-Anderson Nuclear Industry Indemnities Act of 1957, and all renewables subsidies.
- ➢ Per Chapter IV, support women's education worldwide to further reduce population, the one thing that works.
- ➢ Apply lean engineering directly to the research needed in Table 5 by modeling and measuring:
  - o Losses, to improve efficiency
  - o Time and space constraints, to reduce time per unit production. These include equipment size and cycle times
  - o Energy concentration level
  - o Number of parts
  - o Product lifetime/reliability
- ➢ Deploy renewables to replace aging coal plants in the U.S., up to 30%-40% penetration, addressing variability issues through (a) energy storage (via CHP, pumped hydro, CSP, and other thermal storage); (b) higher capacity national transmission networks to increase geographic compensation; and (c) natural gas backup.
- ➢ Eliminate CAFE loopholes to truly achieve a doubling of vehicle miles per gallon.

To pay for these transitions, we should:

- ➢ Appropriately regulate three hidden monopolies:
  - o Company management: Repeal section 162(m) of the U.S. tax code, and rate executive performance
  - o Doctor pay: Regulate doctor pay and open more medical school slots in return for free medical training

- o Education: Improve school funding, teacher standards, and standardize special education requirements
- ➢ Reduce tax code and regulation burdens by applying 5%-10% page count reductions across the board for at least five years.
- ➢ Eliminate subsidy waste: Eliminate corporate tax, agricultural, ethanol, nuclear, fossil fuel and energy subsidies.
- ➢ Change how we think about energy efficiency projects, promoting financing of worthwhile projects to promote economic growth, and monetizing the future benefits of energy consumption in the present.

VII

Insurance Plans

"A bird in the hand is worth two in the bush."
— Latin proverb

The plan outlined above is a first step, espousing long term technological research, substitution of renewables for coal, energy efficiency, economic growth, savings in other areas, subsidy elimination to level the playing field, and changes to EPA rules to level the manufacturing playing field. In the near term, it will only reduce our emissions by 30%-40%, not the 80% reduction called for by the IPCC and global warming scientists. Eighty percent reductions are necessary to stay within a "safe" 2C temperature rise since the Industrial Revolution if climate sensitivity is high.

What if we need reduce emissions farther, sooner?

First, the above plan reduces emissions by 35% relatively quickly, in a few decades, on the way to attaining larger emissions reductions through long term research. But if we need to reduce more, immediately, or buy ourselves more time, there are a few other actions we can take:

Conservation

Earlier, I did not include the notion of limiting usage (spend less, eat less, drive less, and use less) as part of the Economic Law of Pollution

167

because that will slow down our economic growth, and there are structural limits to how much we can reduce fossil fuel use and emissions. But if we *really* need an inexpensive short term reduction, we could go on a fossil fuel emissions diet, and relatively easily cut 5% of emissions without impacting our lives too much (for example, see America's vehicle miles traveled in Figure 19).

When looking at other areas to conserve, and arenas where Americans waste a lot of energy and fossil fuels, two other possibilities arise— trash and food.

Americans put 250 million tons of waste into our landfills in 2012, and recycled about 35% of this[372]. If we further reduce, recycle and re-use this waste, this could generate more short-term emissions savings. For instance, Europeans generally recycle at twice our rate. While recycling produces its own emissions and waste, and recycling economics depends on the recycled materials value and energy content, recycling in general produces lifecycle benefits[373].

The issue of recycling is complicated by a number of market distortions, because landfill costs are appropriately kept artificially low to forestall illegal dumping[374]. To increase recycling rates in the U.S., the next step is to copy the Europeans' extended producer responsibility (EPR) programs, which force manufacturers to take back their goods at the end of life. In the U.S., over 20 states have experimented with such laws, due to opposition at the federal level, especially in regard to electronic waste[375], with some success.

Recycling in the U.S. has improved to the point where the largest component of our trash is not trash, it is food waste[376]. From an 80:20 Pareto perspective, food waste should therefore be our next primary focus. It is estimated that Americans waste 40% of their food[377], for a variety of reasons and at every stage of processing (farm, transportation, store, home, land-fill).

If we apply a lean engineering continuous improvement process of (1) measure/model, (2) reduce waste, (3) compare to theoretical ideal, and (4) improve materials/processes (see Figure 29) to this food waste, we can reduce consumption without harming our lifestyle, and generate emissions savings inexpensively. But we should be careful regarding claims that eliminating meat from our diet is the answer to global warming[378] as espoused by the movie "Cowspiracy". Agricultural emissions are only 9% of our total emissions[379], and beef enteric emissions are only a third of this, or 3%. So eating 25% less beef would only reduce emissions by .7%, and reducing our food waste by 50% will reduce emissions by about 3%.

Overall, then, improved conservation can reduce fossil fuel usage and greenhouse gas emissions by 5%-10%.

Geo-engineering

Geo-engineering is the artificial modification of the Earth's climate, and generally has an unacceptably high risk of unintended consequences. Two classic methods for geo-engineering are to put a lot of dust, mirrors, etc., into the air to reflect sunlight away and cool the Earth, and to fertilize algae blooms to sequester large amounts of CO_2 in the deep ocean.

There are a number of ways to reflect sunlight: (1) Add sulphur, ash, or dust into the air (like a volcano does); (2) put reflecting substances into low-Earth orbits; (3) use a giant reflector in space; or (4) use highly reflecting films on water surfaces[380]. Reflecting sunlight away will cool the Earth and while it is relatively inexpensive, it is also problematic, because it introduces a number of variables and unintended consequences. How much reflection is too much and could we induce an ice age? What happens if the reflection is interrupted suddenly? Who controls the planet's temperature/reflectors? What impact will reflection have on the ozone layer? How would such a scheme impact clouds and their effect on heating and cooling the planet? How can we tell what would actually happen, when our current climate models are unreliable? For all of these reasons, the sunlight reflection geo-engineering option is too risky.

Another geo-engineering idea is to add iron to the ocean (called "iron fertilization"). Iron is a limiting nutrient for phytoplankton growth, so adding it to the ocean creates large phytoplankton blooms. The plankton breathes CO_2 and eventually dies, sinking to the ocean floor where currents trap the CO_2 fixed in the plankton for millennia. However, other studies show that zoo-plankton ocean creatures eat the phytoplankton before they reach a low enough depth where they can be sequestered, so the effect of adding iron is simply to feed the zoo-plankton. Yet other studies show that large diatom plankton blooms are dependent not only on iron availability, but also silicic acid and vitamin B availability, which varies by location.

On the positive side, because iron is a limiting nutrient, half a metric ton of iron can potentially remove as much as 3,000 metric tons of CO_2. But there is also a lot of concern about potential unintended consequences, such as harmful algae blooms, other marine species population booms (jellyfish, zoo-plankton), nutrient depletion of the ocean, etc. It is clear that we do not know enough. And because measurements for the experiments that have been conducted to date are very difficult and/or expensive[381], we may never know enough.

A safer third geo-engineering technique is not well-known and is generally thought to be too expensive[382]. It takes advantage of our ability to produce lots of stuff, piggybacking on how the ocean captures and sequesters CO_2 naturally. About 45% of the CO_2 in the air is absorbed naturally by the ocean, where it forms carboxylic acid, leading to gradual acidification of the ocean. This acid dissolves the carbonate coral reefs; they have reduced in area by 50% since 1985[383], and are projected to disappear entirely by 2050.

If humans add large amounts of pulverized limestone to the ocean, it dissolves to produce carbonate ions that react with atmospheric CO_2

and water to produce two bicarbonate ions, which mineralize on the ocean floor[384]. Adding sacrificial limestone to neutralize natural carbonic acid and reduce coral reef dissolution has not really been investigated enough to tell (a) what the unintended side effects may be, including changes in ocean pH, (b) its effectiveness and speed in removing atmospheric CO_2, (c) what level of dispersion and depth is needed, and (d) how local limestone supersaturation might impact local ecosystems. If 1 mole of limestone equals removing 1 mole of carbon dioxide, then removal of 1 metric ton of CO_2 would require 2.27 metric tons of limestone.

To neutralize our 5,200 million tons of annual CO_2 emissions would therefore require 12,000 million tons of limestone, which is 9 times the quantity of crushed rock (mostly limestone) mined in the U.S. today. This is a stretch to scale up, but is certainly feasible.

Limestone makes up 10% of the total volume of all sedimentary rocks, and originally forms from the precipitation of calcium carbonate from water, so it is located near the coasts, and is not supply limited[385]. This same fact means that transportation costs for this technique are de minimus. One needs to simply dig up limestone on the coast, grind it to the appropriate size, agglomerate it to an appropriate density[386], and drop it into the ocean. By varying porosity and density, agglomerates can float at controlled depths and dissolution rates/distances. Monitored agglomerates can then use ocean currents to float throughout the ocean prior to dissolution to distribute the limestone evenly and freely.

Open pit mining costs are about $5 a metric ton[387], and grinding and agglomeration costs are likely in the $5-$15 range[388], so this is a very inexpensive way to offset our fossil fuel emissions while simultaneously cleaning up ocean acidification naturally. Only a few production plants would be needed worldwide, likely in cold locations in the northern Pacific[xxxviii].

If this speculative technique proves out—and a lot more modeling and testing are needed—this third approach could potentially allow the fossil fuel industry to sell emissions compensated electricity, natural gas, and gasoline, cost only .6% of U.S. GDP, allow full use of all fossil fuel reserves, and counteract ocean acidification. Note, this idea is different from that on www.cquestrate.com (no heat is required), or the ocean alkalinization method evaluated in 2014 by Keller et al[389] (they used $Ca(OH)_2$ rather than limestone at ineffectively low concentration levels).

Summary

In summary, if events prove we have already entered the land of unacceptable consequences and we find that we need to reduce emissions faster than the plan already presented, improved conservation can potentially lower our emissions by a further 5%-10%. Areas of focus include reducing miles traveled, improving recycling rates, and reducing our food waste.

Geo-engineering ideas are for the most part too dangerous, but one idea that needs further investigation is all natural, inexpensive, and potentially effective: dump porous floating limestone into the ocean to disperse, dissolve, and neutralize ocean acidification[390].

[xxxviii] Calcium carbonate is insoluble at the surface of most of the ocean because the ocean's surface is saturated by the presence of calcareous life forms (corals, Foraminifera, etc.) In the deep oceans, there is less calcium carbonate present, and it dissolves readily. The boundary between the two is called the "carbonate compensation depth" (CCD), and it varies depending on pressure, temperature, dissolved CO_2, upwelling, etc. Adding limestone to counteract ocean acidification is more effective at a shallow CCD and/or where fresh water is entering the ocean (i.e. river mouths, glacier melt discharge areas, etc.)

VIII

What You Can Do Locally

"We can begin by doing small things at the local level, like planting community gardens or looking out for our neighbors. That is how change takes place in living systems, not from above but from within, from many local actions occurring simultaneously."
– Grace Lee Boggs

Because of special interests' iron grip on the status quo, implementing many of the plans recommended above will be difficult at the federal level; politicians do not like to repeal popular subsidies. We could ask our politicians to implement offsetting subsidies to make the playing field more even, like the current solar 30% investment tax credit. This will lead to higher levels of market distortion and waste, similar to what we have now—corn ethanol subsidies leading to higher food prices, fossil fuel subsidies not reflecting the true cost of emissions, nuclear subsidies that make taxpayers shoulder risks the market won't, etc.

Given the situation at the federal level, it may be more appropriate for individuals to seek action at state or local levels. From my perspective, the largest emissions sources are (1) my car, (2) my electric power sources (coal and natural gas), and (3) my heating sources (oil, natural gas, heat pump). We do not have a lot of control over these emissions, because they are governed by distant car manufacturers, home builders, and utilities.

The Internet is full of advice about the little things one can do to improve the planet and reduce carbon emissions—plug a leak, plant a tree, keep your tires inflated, use compact fluorescent bulbs[391], pay bills online to save paper, buy food locally, use e-tickets instead of paper[392]—but many of these little things don't work, or use more CO_2, or won't have much of an impact[xxxix].

Instead of an "every little bit helps" approach, lean engineering teaches us to focus on our top emission sources in Pareto fashion. From an individual's perspective, we should therefore:

(a) Buy a higher mpg vehicle (whether hybrid, electric, or gasoline).

(b) Purchase or lease a solar system, or participate in community wind or solar offerings by your utility to obtain non-fossil fuel based electricity.

(c) Provide funding for the research needed shown in Table 5.

(d) Provide political support and money for the above action plan or similar causes.

(e) Talk with your fellow Americans about this plan. Listen. Accept their views. Share your own. In a word, care. And share your caring.

(f) And, finally, focus on eliminating waste in your life, especially food waste, trash waste, and energy waste (turn off light bulbs, turn down thermostats).

With fossil fuel prices increasing due to higher costs of extraction, and with the U.S. dumping more CO_2 into the atmosphere by weight than all of our worldly goods each year, it will take a large group of

---

[xxxix] In the U.S., most paper is already recycled so using e-tickets or paying bills online won't save much. Buying local can actually increase carbon use if your local farmer is farming at a much smaller scale than someone more distant. CFLs, while more efficient than incandescent bulbs, contain mercury (conversely, because they're more efficient, they save on mercury emissions from coal plants).

thoughtful committed citizens[393] to slowly pivot our economy away from fossil fuels and towards future growth.

Please join us in taking the first steps.

This book was written to create an even-handed view of global warming, walking a middle road between media scare stories that sell newspapers and the extreme views of James Hansen, and the denier's view that the global warming is a hoax, is not man-made, and will cost too much. It is the start of a conversation.

The truth is that global warming is real, is man-made, is readily solvable, won't cost much to solve, and we have the time to solve it. But we <u>do</u> need to get started, and activate our market. There are actions we can take.

If you believe this as well, please consider buying five copies of this book, and giving them to people you know. Then talk these ideas over with them -- and help spread the word. All proceeds will go toward the ideas therein.

Appendix A: Glossary & Abbreviations

Anoxic	Waters depleted of oxygen
Anthracite	Highest ranking of coal grades, >85% carbon (dry basis)
ARRA	America Recovery and Reinvestment Act of 2009
Biochar	Charcoal used as a soil amendment
Bituminous	A mid-ranked coal grade, 65%-85% carbon (dry basis)
BOF	Basic oxygen furnace, a type of steel furnace
CAES	Compressed air energy storage
CAFE	Corporate Average Fuel Efficiency standards for automobiles and trucks
Carbon emissions	The CO_2 produced when fossil fuels are burned. Used interchangeably with "fossil fuel emissions" and "CO_2 emissions"
CCGT	Combined cycle gas turbine
CCS	Carbon sequestration and storage
CDIAC	Carbon Dioxide Information Analysis Center, a division of Oakridge National Laboratory
CEO	Chief executive officer
CFC	Chloro-fluorocarbon, the chemical name for a class of refrigerant chemicals that were discovered to react with ozone in the upper atmosphere, and have since been banned
CHP	Combined heat and power. Where electricity is generated and the waste heat is effectively used, typically in manufacturing or to heat building(s)
CO	Carbon monoxide
CO_2	Carbon dioxide, an invisible gas that is a part of our atmosphere

CO_2 emissions	The CO_2 produced when fossil fuels are burned. Used interchangeably with "fossil fuel emissions" and "carbon emissions"
Co-generation	Synonymous with CHP
CSP	Concentrating solar power, an alternative solar technology.
DDT	Dichlorodiphenyltrichloroethane, an insecticide
DI	Direct injection, where fuel is injected directly into a cylinder rather than a port in an internal combustion engine, for higher efficiency
DMAIC	Define, Measure, Analyze, Improve, and Control
EAF	Electric arc furnace
EIA	U.S. Energy Information Administration, which provides independent energy statistics
Electrolyzer	Machine that splits water into hydrogen (H_2) and oxygen (O_2)
Enteric Fermentation	Digestive process in cows that produces methane, a greenhouse gas
EPA	U.S. Environmental Protection Agency
EROEI	Energy returned on energy invested
FOIA	Freedom of Information Act
Fossil fuel emissions	The CO_2 produced when fossil fuels are burned. Used interchangeably with "carbon emissions" and "CO_2 emissions"
GDP	Gross Domestic Product, a measure of the dollar value of all goods and services produced in a year, i.e. the size of the economy
Geo-engineering	Artificial modification of Earth's climate, usually through either solar radiation management and carbon dioxide removal
GRACE	Gravity Recovery and Climate Experiment—twin satellites launched in 2002 that measure Earth's gravitational field
Gt	Giga ton, 1,000,000,000 metric tons

H$_2$	Hydrogen
Humus	Organic component of soil that contains high amounts of carbon that don't break down into CO$_2$
ICE	Internal combustion engine, i.e. an automobile engine
IPCC	Intergovernmental Panel on Climate Change. Established in 1988, a UN group of scientists that reports on climate change to worldwide governments.
ITC	Investment tax credit
kWh	kilo-watt-hour, a measure of power
Kyoto	The Kyoto Protocol is a 1997 international treaty to limit greenhouse gas emissions from 2008-2012.
LEO	Low earth orbit
Lignite	The lowest ranking of coal grades, less than 60% carbon (dry basis)
Methane	CH4, a greenhouse gas that is 20 times more potent than CO2
MW	Mega watt
N$_2$	Nitrogen
NaOH	Sodium hydroxide is also known as lye, or caustic soda.
NCLB	No Child Left Behind Act of 2002
NIMBY	Not in my backyard
NOAA	National Oceanic and Atmospheric Administration, a federal agency focused on the condition of the oceans and the atmosphere
NO$_x$	Nitrogen oxides, a component of smog
O$_2$	Oxygen
OECD	Organization for Economic Cooperation and Development

Pareto	A rule of thumb: 80% of the problem is due to 20% of the causes. Named after Vilfredo Pareto, an Italian engineer and economist.
PCM	Phase change material
PgC	Petagrams of carbon = 1 Giga-metric tons of carbon
PIRG	Public interest research group
ppm	Parts per million
RCP	Representative concentration pathways; CO_2 scenarios by the IPCC
SCCT	Single cycle combustion turbine
SO$_2$	Sulfur dioxide, a component of acid rain
Sub-bituminous	The second lowest ranked coal grade, 60%-70% carbon (dry basis)
Terra preta	A black carbon-heavy fertile soil found in the Amazon
VAT	Value-added tax
VMT	Vehicle miles traveled
WIPP	Waste isolation pilot plant, in New Mexico
WTO	World Trade Organization
WWI	World War I
WWII	World War II

Appendix B: Top 4 through 10 Emission Sources

In Chapter IV, the Economic Law of Pollution was applied to our top three emission sources in 80:20 Pareto fashion. This appendix follows the same format, but examines emission sources four through 10. Because these sources are each a maximum of 5% of our total emissions, the emissions reduction potential for each source are likely low, in the range of a few percent at best. Nevertheless, these sources will become our primary focus once coal and automotive emissions are reduced. In addition, because our industrial emissions have been moved overseas to some extent as our economy has shifted from manufacturing toward services, our industrial emissions have been largely exported. As a result, U.S. industrial emissions have likely been underestimated.

Industrial Emissions

Industrial CO_2 and CH_4 emissions are an integral part of manufacturing processes and materials that support our modern lifestyles. These include steel, aluminum, cement, limestone, and glass manufacturing, and high temperature processing (which generally uses natural gas, the least expensive way to generate heat).

Iron and Steel Production, and Metallurgical Coke Production

Metallurgical coke is used to produce iron or pig iron feedstock from raw iron ore. It is produced by heating coking coal in a low oxygen environment, which drives off the volatile components of the coking coal. A number of byproducts are produced and used in a variety of applications:

Iron is produced by reducing iron oxide with metallurgical coke in a blast furnace. CO and CO_2 are produced with the iron, and the CO is converted to CO_2[394].

Steel is produced by mixing pig iron, scrap steel, natural gas, fuel oil, fluxes (limestone, dolomite) in a furnace. In a basic oxygen furnace (BOF), the carbon in iron and steel scrap combines with high purity oxygen to reduce the carbon content of the metal as desired. CO_2 emissions occur throughout the reduction process. In an electric arc furnace (EAF), an alternative process, the carbon electrodes are consumed and form CO_2 at much lower levels.

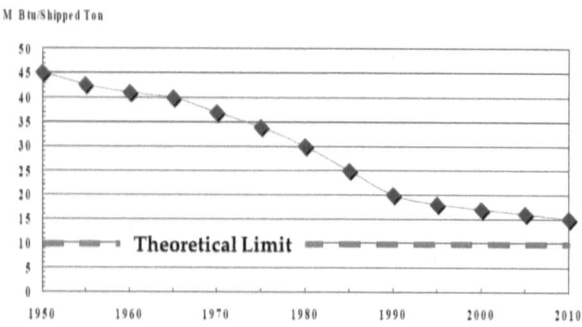

Figure 54. U.S. Steel Industry Energy Efficiency[395]

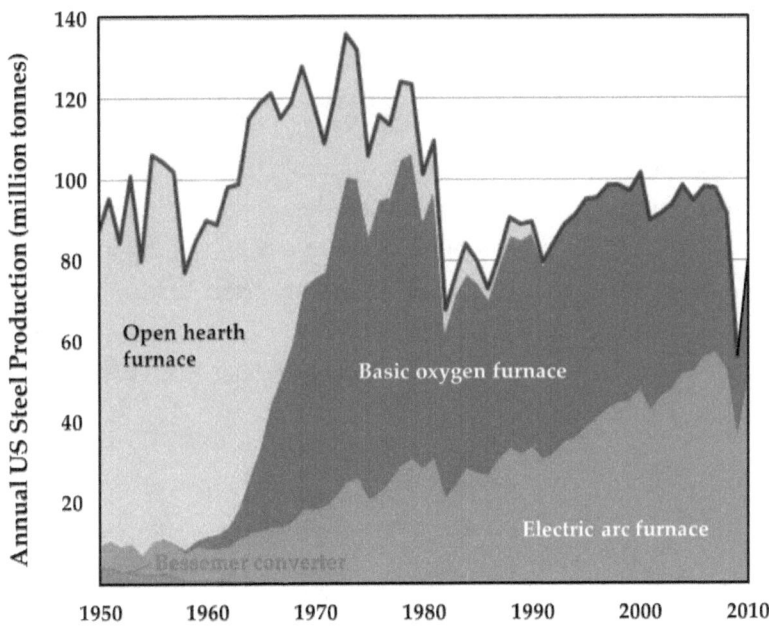

Figure 55. U.S. Steel Industry Furnace Type History[396]

CO_2 is also generated from burning natural gas to produce heat throughout the entire process. Overall, initial Bessemer blast furnaces in 1860 produced 10 tons of pig iron a day, with no energy conservation measures. In 1910, the open hearth technology produced 500 tons of pig iron a day, with exhaust gases preheating entering fuel and air. Today, the basic oxygen furnace and electric arc furnace produce over 150,000 tons per day in the U.S., with a number of energy saving measures further reducing energy per metric ton, as shown in Figure 55[397]. In addition, in 2011, 92% of steel was recycled in the U.S.[398], dramatically reducing the energy/metric ton for this critical material.

As seen in Figure 54, actual energy usage is approximately three times the theoretical limit. Given losses throughout the system, a reasonable goal might be twice the theoretical limit, or a 33% reduction from today. This can be achieved through continuous improvement, as the industry has improved efficiency about 1% per year for decades.

However, to achieve this level of improvement in less than 30 years, the industry will likely need to change its process paradigm.

The electric arc furnace has much lower emissions than the basic oxygen furnace, so continued switching to this process[399], and potentially elimination of carbon electrodes, could be one way to attain this reduction. Another potential way is to use H_2 reduction instead of using carbon reduction. This could take the form of using the H_2 in natural gas rather than coke or coal, or H_2 by electrolysis. A third way could be to use biomass, such as charcoal derived from "steady state" grown forests.

Mandating exclusively using electric arc furnaces in steelmaking in the U.S., which have much lower emissions than basic oxygen furnaces, would have a low emissions impact (about .1%-.2%) for a relatively low cost. Tradeoffs, which need careful evaluation, include potential higher steel costs for some grades, copper recycling impacts, and others[400].

Cement and Lime Production

Calcium carbonate ($CaCO_3$) is heated at 1,450C to form lime (CaO) and CO_2, in a process known as calcining. Lime is combined with silica, cooled, and is mixed with a small amount of gypsum and other materials (slags, etc.) to make Portland cement. The basic process requires heat, which emits CO_2 when natural gas or coal is burned, and produces CO_2 through the $CaCO_3 \rightarrow CaO + CO_2$ formation reaction. While the cement industry continues to improve efficiency through use of more efficient electric motors and other optimizations, there is a limit to what can be improved. In general, 1 metric ton of Portland cement produces about 1 metric ton of CO_2, when including CO_2 from the reaction, the process heat, and the energy to grind the clinker[401].

Cement cannot be effectively recycled, but can be blended with coal fly ash or ground blast furnace slag to reduce emissions slightly (10%-15%).

Cement can also sequester CO_2 by a natural mineralization process, as it re-absorbs about 19% of the CO_2 produced during manufacture over its lifecycle, albeit very slowly (penetrating to a depth of about 1 mm a year). The rate of carbonation depends on the type of cement, quality, environmental conditions (humidity helps), and the permeability of the concrete. Benefits include increased concrete strength; disadvantages include accelerated corrosion of the steel in reinforced concrete and therefore lower structure life[402]. A few companies pursue this avenue, using other materials[403] to make cement that can absorb CO_2, at perhaps higher cost.

In developing countries, which are modernizing at a rapid pace, cement, concrete, and construction are much larger industries with much larger emissions than in the U.S. While cement represents less than 2% of overall emissions in the U.S., it is 5%-10% of worldwide emissions.

There are also materials that can substitute for Portland cement, but may not attain the same strength and stability, and which may require less CO_2 to manufacture. These include lime, plasters/gypsum, mud mortars, and pozzolanas. Pozzolanas include ashes, burnt clays, and siliceous earths (e.g. diatomite); they chemically combine with lime in the presence of water to form a strong cementing material. Each of these alternatives has advantages and disadvantages, including CO_2 emissions, economics, and availability. For example, gypsum works well in dry environments but is not wholly resistant to moisture, and is therefore used for indoor drywall in the U.S., but is not an exterior material. In dry Mediterranean or Middle East countries it can be used externally.

In addition to these well-known alternatives, there is a new class of geo-polymeric cements that use sodium or potassium soluble silicates plus alumino-silicates (clays, slags, fly ash) to replace the calcium carbonate component of Portland cement[404]. Also, Energetically Modified Cement (EMC Cement) is a new patented product that uses a milling process to allow up to 70-30 EMC to Portland cement blends that are as strong as Portland cement, and have a 70% smaller carbon footprint[405]. Both these products look like they may be able to compete with Portland cement on cost, reliability, and performance once they reach sufficient scale and attain market acceptance.

Limestone and Dolomite Use

Limestone, or calcium carbonate ($CaCO_3$), is used in a variety of industrial applications. It produces CO_2 when forming the reaction product lime (CaO), namely: $CaCO_3$ → CaO plus CO_2. Lime/limestone is used in three primary applications: As a flux or purifier in metallurgical furnaces (i.e. steelmaking), as a component of glass (CaO is about 10% of soda-lime glass, which is about 90% of the market), and as the active component of wet scrubbing flue gas desulfurization systems (FGD). Wet scrubbing systems have about 80% of the market.

For the third application above, some alternative processes exist, but these are not as effective or economical as wet scrubbing with limestone.

Natural Gas Systems

The use of natural gas requires field production, processing, pipelines, storage, and distribution. Methane leaks occur at the well, during transitions, at transfer points, from compressor station leakage, and from incomplete combustion during flaring[406].

At each potential leakage point, the natural gas industry loses revenue, so the industry is incentivized to reduce the leaks. Leaks tend to be intermittent (during maintenance operations, abnormal or emergency conditions, etc.), so they are both difficult to measure and likely to be small. However, because methane is 20 times more effective at global warming than CO_2, even small leaks or intermittent should be minimized. The amount of leakage actually occurring, and the potential need for the EPA regulations, is highly controversial.

Gas-fired Residential Heating

Natural gas, home heating oil, wood, electricity, and propane are the primary fuels used to heat homes in the U.S. In 2012, these sources represented 67%, 12%, 6%, 6%, and 12%, respectively[407] of approximately 7% of our total energy use. Because the price of natural gas is low, it is expected that more homes will switch from heating oil to natural gas over the next decade.

Natural gas furnaces are relatively efficient, with most homes in northern parts of the U.S. using 90%-95% efficient furnaces, and 80% efficient furnaces in the rest of the country[408]. Whether it is worth it to upgrade furnace efficiency for a particular home depends on the furnace's age and efficiency, climate, cost of natural gas, and venting configuration. Oil-fired furnaces tend to be about 80% efficient. The efficiency of wood-fired heat varies widely—from 10% for a fireplace to 30%-50% for a typical wood stove to up to 90% for some designs. Electricity efficiency varies by source (coal, natural gas, etc.) but while the heater coil efficiency is close to 100%, power plant efficiency tends to be low, as it is a mixture of about 30% for coal, 60% for gas, more than 90% for nuclear. Propane furnace efficiency varies between 92%-97%.

As mentioned in the last paragraph, more efficient heating equipment is generally available. Whether the upgrade will pay off over the equipment's lifetime depends on many factors, especially the venting system. Higher efficiency condensing furnaces have different venting requirements that can be expensive to change in existing homes.

Nevertheless, mandating sales of higher efficiency furnaces for new construction would be essentially free over the long term, as the long term fuel savings would pay for the relatively small increase in the purchase price of the furnace[409]. However, this mandate has politically been difficult to achieve, and standards have not changed since 1987.

N₂O Agricultural Emissions

Nitrogen oxide, N_2O, or laughing gas, warms the atmosphere 300 times more than carbon dioxide, and has a 120 year lifetime in the atmosphere. Roughly 70% of nitrous oxide emissions come from the soil through the use of synthetic fertilizers. About 5% each stems from manure management, transportation, stationary combustion (factories), and about 10% is released as a byproduct of the industrial production of nitric acid, which is used to make fertilizer and adipic acid, used to make nylon and other synthetics[410].

Increased use of catalytic converters in automobiles, and the use of low NO_x industrial burners, which have been driven by EPA and other regulations, show that NO_x emissions have been reduced by 50% over the last two decades in the U.S. (see Figure 4). This has partially compensated for increased use of nitrogen fertilizer, approximately 25% population growth over that time, and 12% livestock growth over that time.

However, because 30%-50% of worldwide crop yields are attributed to use of natural or synthetic fertilizer, at least one-third of the world's population is currently fed as a result of worldwide 100 million tons of annual synthetic fertilizer use. Slow release coatings, better distribution and improved fertilizing practices can reduce fertilizer emissions, and the fertilizer industry believes that more widespread use of best practice technologies in fertilizer production will also reduce the industry's footprint[411]. Runoff into rivers, lakes, and oceans, and the oxygen death that this runoff produces is a separate pollution issue[412]; however, the presence of this agricultural runoff points to low application efficiency and a high level of waste.

The addition of biochar/charcoal can reduce the amount of fertilizer needed in the soil, but is still too expensive, as discussed earlier.

Aviation and Marine Emissions

CO_2 is produced when jet fuel (in aviation) and distillate diesel and residential fuel oil (in marine transportation) are burned. Jet turbines and aircraft are highly fuel efficient, and Boeing's 787 airliner, with increased use of electricity throughout the plane, is 20% more fuel efficient than the 777 or other similarly sized aircraft[413].

The history of the airliner shows, in Figure 56, how fuel efficiency has tripled since 1955. These innovations are a complex combination of aerodynamic design, engine improvements, and materials improvements. Similarly, most ships use reciprocating diesel engines for high efficiency. However, we may see more liquid natural gas ship engines being used, as these are more efficient and have sharply reduced emissions compared to diesel engines.

Figure 56. Airliner fuel efficiency trends[414]

Methane Emissions

Ruminant animals such as beef and dairy cattle and sheep produce methane as part of their normal digestive processes when microbes in their fore-stomach (the "rumen") ferment the food consumed. Feed quality and type can impact these emissions. Horses, swine, and sheep also produce methane, but in much smaller quantities.

Some have suggested that U.S. consumers should reduce meat consumption to reduce methane emissions from cattle, and to reduce the amount of fertilizer used for feed. However, a look at the U.S. livestock and poultry industry historical record shows that livestock inventory[415], on average, has only increased about 5% between 1997 and 2007, which is one half the population growth of 10% over the same period.

In Australia, there are ruminant species of kangaroos that produce 80% less methane than cows because a different type of bacteria inhabits their rumen. These bacteria are able to produce succinate as a final product of the lignocelluloses degradation, producing small amounts of methane as end product[416]. It may therefore be possible, through microbial engineering, to change the microbiota composition of the rumen in strong methane producers, thereby reducing emissions.

Non-Energy Use of Fuels

Non-energy fuel uses include the manufacturing of plastics, rubber, synthetic fibers, reducing agents for the production of various metals and inorganic products, and the use of products such as lubricants, waxes, and asphalt. CO_2 emissions arise from the manufacture of these products, or sometimes the use of these products, such as solvent outgassing.

Each of these manufacturing processes and uses has its own efficiencies. Manufacturers generally improve their processes to reduce costs over time, and sometimes find product substitutes that reduce emissions. A number of these products enable and support our modern lifestyle, so, aside from continuous process improvement, emissions may be difficult to change.

Appendix C: Improving Innovation

The U.S. government appropriately supports early stage research into energy, including helping to launch fracking, which has produced recent high economic dividends as well as increased our oil supplies. But some programs have been failures—for example, the DOE loan guarantee program was originally created to help bridge the valley of death but has been highly controversial. The valley of death occurs when a new product is developed with high fixed product development expenses, but sales are not yet large enough to support an enterprise. The Solyndra loan guarantee debacle and other failures have been politicized by the Republican Party as examples of wasteful government spending, but the program as a whole is now in the black and has had a number of successes (it helped launch Tesla, for instance). But the program recently issued large loans to the highly mature (i.e. not risky) automobile and nuclear industries, which do not need help bridging the valley of death[417]; so this program is not really fulfilling its original purpose.

Government has a long history of not being able to pick winners and losers very well[418]; and of sometimes successfully helping get technologies to scale[419]. In private markets, which have done a better job, there are essentially four types of investors, each with a different risk/reward profile:

Table 12. Investor Risk/Reward Profiles

	Expected Reward	Financial Model Focus	Types of Risk
Banks	<10% ROI[420]	"Worst case" scenario to estimate debt-service coverage ratios	Execution Risk ("team executes plan")
Equity	10%-20% ROI	"Expected case" scenario to estimate ROI, with downside and upside cases/risks	Engineering and Scale-up Risk ("manufacture-able at scale & low cost?")
Venture Capital Angel Investor	20-times plus	High risk, 90% of projects fail	Invention Risk ("will it work?")

While there are exceptions, this table shows generally the types of risk that investors shoulder, and the commensurately higher reward they expect for taking higher risk. While not shown, in general, government financiers don't like risk at all, because this gets their funding cut if something goes wrong (as Solydra has demonstrated). The problem with the DOE loan guarantee program is that a well-balanced low-risk portfolio means that riskier projects don't get a lot of funding--- which is counter to the program's purpose of mitigating the valley of death. Also, the program creates a morale hazard for "riskier project" managers because the taxpayers bear the risk-- so these managers don't focus on reducing risk enough.

Because of these market distortions, the loan guarantee program should be replaced by investment risk reduction assistance. This takes the form of funding projects that reduce investment risk directly. The Advanced Research Projects Agency- Energy (ARPA-E) is a good start for addressing "Will it Work?" Invention Risk. More support is needed with regard to addressing scale-up and acceleration of time to market.

America knows how to innovate. Corporations have long employed a product development process that uses "stage gates[421]". At each stage, prospective research projects and new products must pass internal hurdles. The more competition between ideas at each stage, and the more killing off (or redirecting) of bad projects, the more successful the R&D program[422]. Federal R&D assistance attempts to mimic this process[423], but money gets wasted because fund managers aren't strict enough.

The more planning at early stages, the easier it is to get through later stages[xl]. We are now able to model performance/technical, cost, reliability, supply chain, manufacturing process, operational, legal/regulatory, and market risks at much earlier stages than before, and modeling can help focus research on the key variables that can reduce these risks. In addition, we can do real world testing more easily and quickly with 3D prototyping and printing.

Silicon Valley specializes in these types of support, and many areas of the world are attempting to mimic its "Center of Excellence"/economic cluster for entrepreneurs, which includes assistance with company financing, business incubation, and scale-up.

Private markets will generally provide funding if risks are low relative to expectations within one of the four risk levels: bank lending, equity lending, venture capital, and angel funding. To get to scale, investment risks include performance/technical, cost, reliability, supply chain, manufacturing, operations, and market risks. And a lean engineering approach of "Model/Measure, Reduce Waste, Compare to Ideal, and

---

[xl] This is the philosophy behind the highly successful "Design for Manufacture and Assembly" approach to product development.

Improve" (Figure 29, page 81, MRCI) can be applied to each of these risks.

In summary, our innovation engine works well, but could be augmented and accelerated by research programs that focus on reducing investment risk by applying the MRCI model directly.

Appendix D: Other Barriers to Action

Throughout this work, a number of barriers are cited that have stymied progress on de-fossilizing our economy or solving global warming. This chapter enumerates other barriers, and how the action plan above circumvents them.

Freeloading
From a geographical perspective, global warming costs are national, and the benefits are global. That creates a strong incentive for free riding. One barrier to action from the U.S. perspective is that if the U.S. incurs costs, other countries—particularly developing countries— will free ride on our efforts, reducing our competitiveness, and impoverishing us at our own expense. Economist Nordhaus states, "Countries have strong incentives to create ambitious goals and then ignore them. When national economic interests collide with international agreements, there will be a temptation to shirk, dissemble, and withdraw."[424]

We have seen this as the U.S. abandoned Kyoto in 1997, and annual UN climate change conferences—Cancun, Durban, Doha, Warsaw, to name a few, have been largely futile with regard to meaningful CO_2 reductions. William Nordhaus also shows that for carbon tax or other schemes to be effective worldwide, they should be applied uniformly, or polluting industries will shift operations to where carbon taxes are lowest, defeating their purpose.

Both of these issues can been addressed by applying "EPA violation" tariffs on imports that do not meet EPA air, water, toxic substance, or waste pollution standards, and using the funds to support the EPA equivalent in foreign countries for direct pollution controls, strategies, etc. This makes it harder to avoid environmental regulations, and when CO_2 fines are included as part of EPA air pollution standards, will keep other countries from free riding on our efforts. If countries want to sell products in our markets, they should comply with EPA standards.

Climate Justice

If we look at who has produced the most carbon emissions, exceeding the Earth's carbon sinks to the point where CO_2 in the atmosphere has risen by 40% since the Industrial Revolution, then Europe and the United States have produced the most carbon emissions cumulatively, despite being overtaken on an annual basis by China and India in recent years. Should developing countries pay for our past profligacy? Should the developing nations not be allowed to take advantage of their own fossil resources to boost themselves to prosperity as we have done, so that they have the resources to start dealing with the problem? Should they not follow the same development path? Assigning fault for CO_2 pollution geographically touches on international "climate justice" with respect to current and past emissions.

Climate justice—who should pay historically—is complicated. Record-keeping on emissions since the 1700s varies considerably worldwide, and is inadequate for exact determinations of who might be at fault under a "polluter pays" legal framework. In addition to data inadequacy, other controversies include:

(a) Are cumulative CO_2 emissions what should govern? Why not CO_2 gases in our current atmosphere or contribution to temperature increase?

(b) Should land-use considerations be included? If so, deforestation over the last three centuries dramatically increases emissions from tropical countries, such as Brazil and Indonesia.

(c) What starting year should be used?

(d) Should the CO_2 equivalent of non-CO_2 gases be included?

In short, the developing nations' position is that the global warming problem is primarily the developed nations' problem because we caused most of the emissions. The U.S. position, espoused by both Republicans[425] and Democrats[426], is that this burden should be shared by all.

Under a "polluter pays" legal framework, of course, the U.S. is responsible for its past emissions despite the above measurement difficulties[427]. We should therefore lead the charge when it comes to decarbonizing our economy as well as paying for cleanup. This will also improve economic growth by reducing the impact of the recent 5% GDP increase in the costs of using fossil fuels.

Poverty Reduction

Five-sixths of the world's population lives in poverty, relative to the developed world. As these people reduce poverty and demand western levels of energy consumption, emissions growth will accelerate.

As poor countries economically develop and shift from rural economies toward manufacturing in cities, they become less vulnerable to climate change because they have more resources. Population is reduced sharply, reducing emissions; and CO_2 emissions per capita[428] rises because they use the least expensive fuel source, fossil fuels, to power themselves out of poverty.

This is all well and good, except that coal, oil, and natural gas extraction costs are increasing worldwide (with fracking creating an exception with regard to natural gas, but the applicability of this method depends on both infrastructure availability and underlying geology). As fossil fuel extraction costs increase, and inexpensive sources in the Middle East dry up, countries will be less and less able to use fossil fuels to expand their net energy per capita to exit poverty. They need our help finding new sources of inexpensive energy.

Jevon's Paradox

Jevon's Paradox states that as energy efficiency increases, the cost of energy falls, resulting in the rate of consumption increasing due to higher demand. This higher consumption offsets the reduction in emissions, so that energy efficiency cannot reduce emissions.

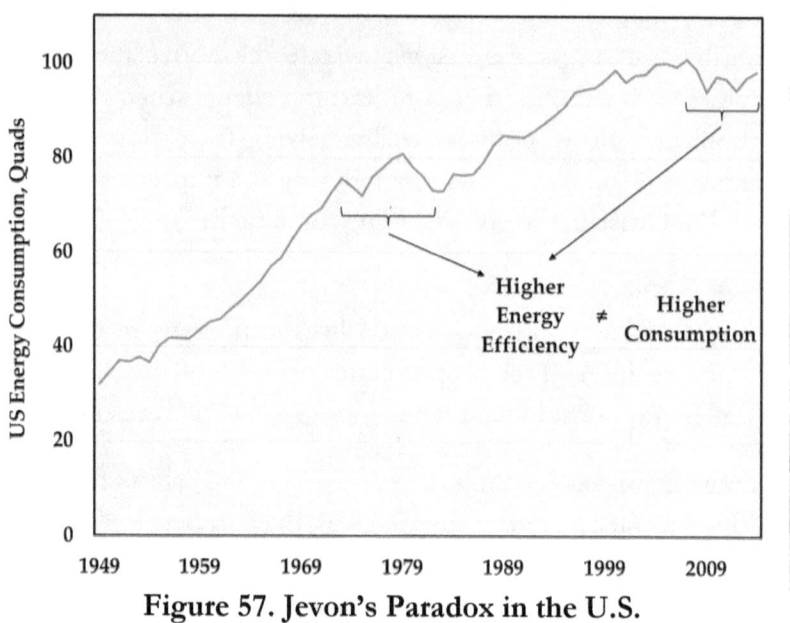

Figure 57. Jevon's Paradox in the U.S.

However, Figure 57 shows that U.S. energy consumption decreased when energy efficiency increased, during periods of high oil prices in the 1970s and again in 2005+. Jevon's Paradox simply doesn't apply to the U.S. economy.

Exponential Growth = Unaffordable CO_2 Mitigation Costs

In 1972, a few MIT researchers created a "system dynamics theory" computer model to analyze the long term causes and consequences of growth in world population and the world's material economy. Based on the model, they published a book called "Limits to Growth" that posited that a limited pool of resources combined with exponential growth will lead to collapse sometime in the 21st century, beyond

2015[429]. Interestingly, their population predictions have been confirmed, but projections for resource price and availability were spectacularly wrong, as we developed new technologies to find and extract resources.

The authors also posit that exponential growth in both economy and population implies that CO_2 mitigation is ultimately unaffordable. Their logic is as follows:

Our society—the government, all companies, and us as individuals—pursues economic growth as a primary method of increasing our welfare. Capitalist society, with a positive feedback loop provided by capital[xli], has an inherent structure that produces exponential growth; in addition, as shown in Figure 23, population growth is exponential. Our food supply, materials, energy use, and economy have all been growing exponentially since the Industrial Revolution, as has CO_2 pollution. As we have improved our labor productivity, we have all gotten exponentially richer (even if this wealth is not distributed evenly).

Then, the authors posit that pollution costs will also rise exponentially, making CO_2 mitigation unaffordable:

"It is usually inexpensive to remove 50% of pollution; more expensive to remove 80%; and very expensive/prohibitive to reduce the last 1%. By implication, it may therefore be affordable to cut pollutants per car in half; but then if the number of cars doubles due to exponential growth, the pollutants per car must go down by 25% again … and at some point it stops being true that growth will work." [430]

---

xli In modern economies, we set aside funds to replace equipment as it wears out (called depreciation); this, plus extra funds devoted to improving our processes/equipment/factories/offices, provides a positive feedback loop. Growth leads to more resources to promote growth.

The logic espoused by the "Limits to Growth" school is erroneous for three reasons:

(1) Population growth is reducing, not increasing exponentially
(2) The assumption of non-linear cost increases as limits are approached neglect to take into account the Economic Law of Pollution—the option of substitution, while not fully practical today, can be found through further research and technical innovation
(3) In general, the world has been technologically innovating its way around limits—from agriculture and fish farms to energy, to population; there is no reason not to expect us to find technological solutions to global warming.

The "Limits to Growth" crowd continues to think that we are running out of resources, and number of authors claim that the end of growth is coming[431]. And they may have a point relative to Moore's Law as we approach the physical atomic limits of silicon. This will reduce computer industry productivity as we search for a replacement "It can't continue forever. The nature of exponentials is that you push them out and eventually disaster happens." (Gordon Moore)[432]

But the "Limits to Growth" crowd, from Malthus onward, has repeatedly turned out to be wrong in the face of lean engineering efficiency, scale, and paradigm innovation. The world's oceans, previously thought to provide an unlimited supply of fish, are now relatively exhausted, to the point where 42% of fish in 2012 were procured through fish farming[433], which has allowed us to exceed the limited supply of natural fish. In agriculture, the use of fertilizers has expanded food production per acre by over 48%, allowing more population worldwide than would otherwise be possible[434].

Lean engineering teaches that exponential growth—i.e. continuous improvement—is sustainable, and can reduce poverty, promote climate justice, and address freeloading problems relative to global warming.

Appendix E: Global Warming Feedbacks

Despite the imperfections of climate science, we know:

➤ Carbon flows between land, oceans, and air (for example, only 56% of the CO_2 we emit to the air stays in the air[435])

➤ Gases other than CO_2 impact the climate (nitrous oxide, methane, CFCs, water vapor, etc.)

➤ Smog, dust, clouds, and other pollutants and aerosols impact atmospheric temperatures

➤ Human CO_2 emissions keep climbing, in lockstep with economic activity and population growth

➤ Temperature increasing "positive" feedback effects include:

    ○ Dark areas previously covered by snow. As more snow melts, the dark earth exposed absorbs more sunlight, speeding up warming and melting the snow faster. We can see this feedback mechanism operating when snow melts on our driveways before it melts on the surrounding lawn.

    ○ The greenhouse effect. CO_2 is nearly transparent to solar radiation, but partially opaque to infrared (IR) radiation emitted by the earth. Thus solar radiation can pass through the atmosphere to hit the Earth's surface. The surface warms, and emits IR radiation. Some of this energy is absorbed and re-radiated by the upper atmosphere's CO_2 back toward the Earth's surface. More CO_2 increases this absorption and re-emission, warming the Earth[436]. This blanket of .04% CO_2[xlii] actually increases global temperature by about 33C relative to their absence[437], so these trace gases are critical for life on Earth.

    ○ CO_2 solubility in the ocean. When the ocean warms up, it releases CO_2 into the air.

[xlii] CO_2 comprises 400 part per million (ppm) = .04% of our atmosphere.

o Methane has 20 times the global warming effect of CO_2, but stays in the atmosphere for a much shorter time (10 years vs. ~100s of years for CO_2). If the methane stored in methane hydrate formations on the ocean floor is released, as occurred during the Permian Extinction 250 million years ago, global temperatures might run away to Venus-like conditions. However, this is highly unlikely. The shallow Arctic shelf, which is emitting methane now, may contain less than 1% of the world's gas hydrates. About 3.5% of hydrates occur in the upper edge of stability, 1,000-1,600 feet down; at these depths, warming released methane would likely oxidize before reaching the surface. And 95% of hydrates occur at depths greater than 3,000 feet, so any methane released would be oxidized or re-trapped in sediments prior to reaching the surface[438].

o Methane is also produced as the permafrost melts in areas where newly warmed decaying matter is not exposed to oxygen—i.e. Arctic and Antarctic swamps, and wetlands. Wetlands cover about 6% of Earth's land area, down from 12% due to humans draining swamps for development, but so far paleontological research shows that large methane releases during previous warming periods may not have happened[439], so while this contributes to a positive feedback loop, it does not appear to be driving new levels of warming at the moment.

➤ Temperature decreasing "negative" feedbacks include:

o CO_2 fertilization: When CO_2 levels increase, plant growth can increase by 20%-25%[440]. This results in a one-time uptake of carbon associated with the sequestration of this additional 20%-25% growth[441].

o Longer growing seasons: Dead plants have been frozen in the permafrost, and CO_2 will be released as they rot when the permafrost melts. Conversely, longer growing seasons at the poles cause new forests to grow, which is a carbon sink. Currently the poles are a sink rather than a source[442].

o Volcanic ash and other aerosols: These can block sunlight, reducing the Earth's temperature.

o CO_2 ocean solubility: Non-CO_2 saturated cold water can absorb CO_2 (turning the ocean more acidic), lowering atmospheric CO_2 levels. As glaciers and Greenland's ice melts, it allows more CO_2 to be absorbed by the ocean.

o Weathering: Over long periods of time, CO_2 in the air precipitates into weak carbonic acid and weathers rocks that precipitate carbonates onto the ocean floor[443]. This process takes many thousands of years to remove significant amounts of CO_2 from the air.

➢ Impacts are localized and highly variable, including precipitation impacts (both drying out/desertification, and getting wet/tropical). There will be further surprises regarding both positive and negative feedbacks, especially regarding impacts in specific locations.

Appendix F: EROEI

Because fossil fuels are the energy source for our modern economy, we also need to consider the implications of higher fossil fuel costs on an "energy basis" (as opposed to cost).

In 1959, anthropologist Leslie White showed that the level of wealth and culture in a society is determined by the net energy available per person. A corollary is that economic growth requires increasing net energy available per person. We have seen this in China's recent growth, fueled by four-plus coal plants built per day, and in the U.S., where exponential productivity improvements, efficiency improvements, and the use of fossil fuels provide excess net energy. When net energy available per person decreases, our economy shrinks—as Figure 11 shows.

The still-valid point of the "Peak Oil" argument is not that oil is running out, but that *cheap oil* is running out. Technology advances will continue to increase access to resources, but the cost of this access will continue to rise. In the future, it will not make economic sense to extract the oil or the coal (and later the gas), and alternatives will be needed.

While there is plenty of oil today, conventional resources are drying up as evidenced by few announcements of new conventional fossil fuel field discoveries. Unconventional resources—oil recovered from fracking, oil tar sands, tight oil, etc.—are being exploited (and announced). The U.S. has become another Saudi Arabia relative to fracking tight-oil production. However, these unconventional sources incur higher costs than conventional resources, as either heat (to make the tar flow) or pressure/water/fracking is needed to extract and process the oil. A method of measuring these costs is called "Energy returned on energy invested", or EROEI.

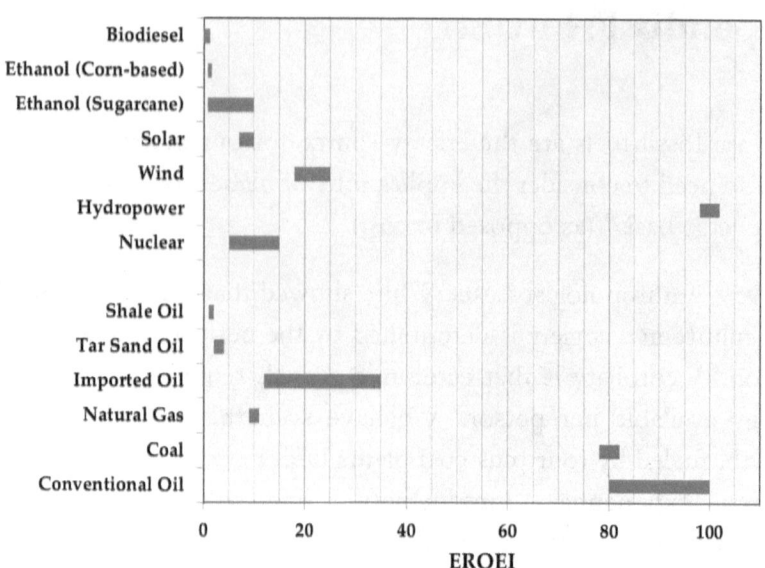

Figure 58. Energy Return On Energy Invested (EROEI)[444]

EROEI equals "Energy output" divided by "Energy input," and is therefore a measure of the efficiency of extracting an energy source; the energy leftover is then available to run our modern economy. For example, for gasoline, EROEI is the energy in a gallon of gasoline divided by all the energy required to make the gasoline, including oil exploration, drilling, and refining[xliii]. If EROEI is 100, we get 100 barrels of conventional oil for 1 barrel of oil energy spent to extract and process these 100 barrels. If EROEI is 10, we get 10 barrels of oil for 1 barrel of oil energy spent. If EROEI is 1, we get no excess energy and might as well have not bothered, except maybe to use the oil as a feedstock for a chemical process.

Another way to represent this is called the "net energy cliff," which illustrates that as EROI approaches 1:1, the ratio of energy gained

---

[xliii] Note, as well, there are several methods of calculating the inputs, depending on what is or is not included. See www.wikipedia.org: Energy Return on Energy Invested for further information.

(dark red/grey) to energy used (green/light grey) decreases exponentially:

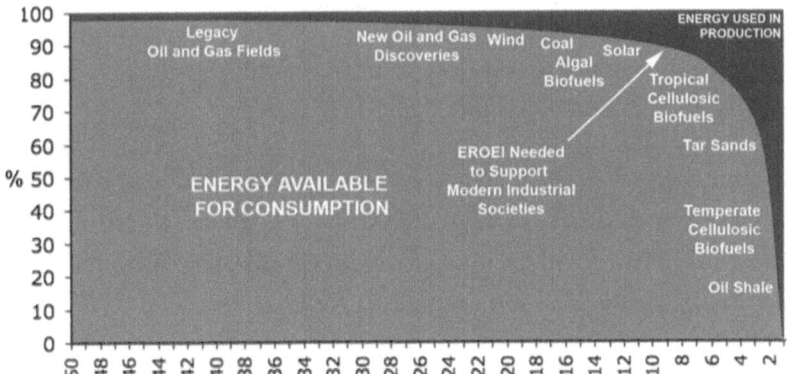

ENERGY RETURN ON ENERGY INVESTED (EROEI)

Figure 59. The "Net Energy Cliff" [445]

Some key points of these last two graphs include[446] [447]:

- Conventional oil had an EROEI of 80-100 a century ago
- This has dwindled, so our average EROEI for U.S. oil today is about 10, a blend of conventional and unconventional resources
- Unconventional tar sands and shale oil have EROEI of about 2-5
- Coal has an EROEI of about 80, so is inexpensive
- Natural gas has an EROEI of 10, similar to oil
- Nuclear has an EROEI of 5-15
- Wind has an EROEI of 18
- Solar has an EROEI of 7-10
- Corn ethanol has an EROEI of 1.2

Given that a bare minimum EROI for a modern economy[448] is estimated to be about 5, the implications of these graphs are clear: the "black gold" of oil—either shale oil or tar sands—cannot sustain our transportation sector from a net energy perspective. Inexpensive oil is running out, and this is a serious drain on our economy. Even the tight oil and shale oil boom that fracking technology has enabled may give us only a short reprieve—the Energy Information Administration projects that our percentage of oil imports will dip to 34% by 2020, then rise to 45% by 2040 in its Reference case[449].

In summary, while inexpensive natural gas and coal are likely to be plentiful for another two generations, inexpensive oil has been depleted. This means that our wealth and economic growth will shrink, as natural gas cannot readily substitute for oil.

But, wait, don't we have 200 years or more of coal available? And even more fossil fuels available worldwide?

In fact, all coal-producing governments in the world have in the past vastly overestimated economically viable coal reserves until falling production forces them to say "oops." The U.S. is no exception, with anthracite production peaking in Pennsylvania in 1915. In Germany, the Ruhr valley peaked in the 1950s. In the UK, coal peaked in 1913; etc.[450] Interestingly, M. King Hubbert applied coal's historical peaking behavior to accurately estimate oil's peak in 1970 as shown in Figure 10.

Similar to coal estimates, most oil producing countries' estimates of reserves are suspect and unreliable (as investors worldwide can attest).

U.S. government estimates of coal reserves follow the same pattern of overestimation[451], and academic studies predict that 90% of economically recoverable coal will be exhausted by 2070[35]. EIA estimates of how long oil and gas reserves will last at current usage rates show 2070, a little over two generations, which matches academic estimates[452].

Appendix G: End Notes

[1] Cover Image Photos attribution. The rest of the photos were taken by US government employees and are not protected by copyright: Hurricane_Rita_Peak-NOAA-Public-Domain-Wikipedia-Commons.jpg, 793px-Hurricane_Katrina_August_28_2005_ NASA, satellite photo of Hurricane Katrina in the North Atlantic, noaa.gov; Chenaga Glacier.jpg, Chenega Glacier located in the Chugach National Forest, Chugach Mountains, Prince William Sound, 17 June 2004, US Fish and Wildlife Services archive; FEMA - 36445 - Aerial of flooding in Missouri.jpg: Jocelyn Augustino, St. Louis, MO, June 20, 2008 -- Flood waters impact areas west of St. Louis, FEMA Photo Library; DDR4241greategrets102815BK2284_medium.jpg: Bill Koplitz, Columbia SC, Lake Elizabeth after part of its Levee broke, Oct 28 2015, FEMA Photo Library; 50642_medium.jpg: Prattsville, NY 2011, Flooding near this river bank resulted in soil erosion and damage to surrounding buildings. Rains caused by tropical storm Irene brought significant flooding to the area

[2] See "A Bomber An Hour", www.strategosinc.com/Willow_run.htm

[3] Inventory held to showcase products, such as in a bookstore, helps to support sales, but is still wasteful from a manufacturing point of view, as the print and ship on demand model for books is showing. This list is an ideal one strives toward, but final goods inventory may still be "necessary" for many products.

[4] Address to the National Editorial Association, Jamestown, Virginia, June 10, 1907.

[5] Emissions: Carbon Dioxide Information Analysis Center (CDIAC), www.cdiac.ornl.gov. Atmospheric emissions are Mauna Loa(1950+), and Law Dome ice cores prior to 1950. GDP per Capital: www.worldbank.org

[6] Jennifer Welsh, "Man Entered the Kitchen 1.9 Million Years Ago", Aug 22, 2011, www.livescience.com

[7] Richard Heinberg, "Snake Oil: How Fracking's False Promise of Plenty Imperils Our Future", 2013, Chapter 6.

[8] www.cdiac.ornl.gov

[9] So far there is evidence of relative decoupling of the economy and fossil fuels in some areas, but not absolute decoupling. See, for example, Mark Burton, "The Decoupling Debate: Can Economic Growth Really Continue Without Emission Increases?", www.resilience.org, Oct 2015.

[10] *Massachusetts et al v. Environmental Protection Agency et al*, decided 4/2/2007, US Supreme Court Docket # 05-1120; and *Utility Air Regulatory Group v. Environmental Protection Agency et al*, decided 6/23/14, US Supreme Court Docket No 12-1146.

[11] Jonathan Adler, "Supreme Court puts the brakes on the EPA's Clean Power Plan", Washington Post, Feb 9, 2016.

[12] See L. Hartwell Allen Jr, Jeff Baker, and Ken Boote "The CO_2 fertilization effect: higher carbohydrate production and retention as biomass and seed yield",

[13] Adam Welz, "The Surprising Role of CO_2 in Changes on the African Savanna", Jun 2013, Yale Environment 360.

[14] Coco Ballantyne, "Strange but True: Drinking Too Much Water Can Kill", June 21, 2007, Scientific American

[15] Dana Dovey, "Water Intoxication: Just How Much $H2O$ Does It Take To Kill A Person?", www.medicaldaily.com

[16] Patel et al, "Oxygen Toxicity", JIACM 2003, Vol 4, #3, p 234-7.

[17] Norhaus, William, "The Climate Casino: Risk, Uncertainty, and Economics for a Warming World", Yale University Press, 2013, p 8 and 17.

[18] William Nordhaus, "The Climate Casino: Risk, Uncertainty, and Economics for a Warming World", Yale University Press, 2013, Chapter 23

[19] Oreskes, Naomi, and Conway, Erik. "Merchants of Doubt", Bloomsbury, New York, 2010.

[20] Norhaus, William, "The Climate Casino: Risk, Uncertainty, and Economics for a Warming World", Yale University Press, 2013, pp 287.

[21] "Psychology and Global Climate Change: Addressing a Multi-faceted Phenomenon and Set of Challenges", a report by the American Psychological Association's Task Force on the Interface Between Psychology and Global Climate Change", Swim et al, 2009.

[22] http://changingminds.org/disciplines/change_management/psychology_change/psychology_change.htm

[23] http://www.gallup.com/poll/168620/one-four-solidly-skeptical-global-warming.aspx

[24] http://www.gallup.com/poll/167879/not-global-warming-serious-threat.aspx

[25] Leiserowitz, et al, "Climate Change in the American Mind", March 2015, George Mason University.

[26] Interestingly, Rachel Carson got a lot wrong, especially the axiom "the dose makes the poison" of modern toxicology.-- see Henry I. Miller and Gregory Conko, "Rachel Carson's Deadly Fantasies", Forbes, Sep 5, 2012.

[27] See www.epa.gov/superfund; www.hanford.gov/page.cfm/HanfordCleanup; John Berg, "The cleanup of Boston Harbor was surprisingly triumphant", Aug 1 2004, Commonwealth Politics, Ideas, & Civic Life in Massachusetts; ozonewatch.gsfc.nasa.gov; www.epa.gov/acidrain/; ...

[28] Benjamin Franklin, Poor Richard's Almanack, 1758.

[29] http://www.epa.gov/ttn/chief/trends/ file national_tier1_caps .xlsx

[30] Emily Atkin, "Why This New Study on Artic Permafrost Is So Scary", Climate Progress, Apr 8 2015.

[31] Sam Carana, "Horrific Methane Eruptions in East Siberia Sea", Aug 31, 2014, Artic News

[32] Justin Gillis and Kenneth Chang, "Scientists Warn of Rising Oceans from Polar Melt", New York Times, May 2014

[33] Chris Mooney, "The melting of Antarctica was already really bad. It just got worse.", Energy and Environment, Mar 16, 2015.

[34] James Hansen, May 2015.

[35] www.agcensus.usda.gov and www.census.gov

[36] noaa.gov, http://www.nws.noaa.gov/om/hazstats.shtml.

[37] New York Times editorial, 11/27/12, "Hurricane Sandy's Rising Costs"

[38] http://www.nhc.noaa.gov/data/tcr/AL182012_Sandy.pdf; note that NOAA official statistics at http://www.nws.noaa.gov/om/hazstats.shtml do not reflect the figures in this report, in part because damages can be spread out over time, and/or lead to renewed economic activity (i.e. reconstruction). See http://www.esa.doc.gov/sites/default/files/sandyfinal101713.pdf for a more complete picture.

[39] noaa.gov

[40] Laurens Bouwer, "Have Disaster Losses Increased Due to Anthropogenic Climate Change?", Bull. Amer. Meteor. Soc, V92, #39-46, 2011.

[41] "TyphoonHaiyan - RW Updates". United Nations Office for the Coordination of Humanitarian Affairs. December 28, 2013. *Philippines: Hundreds of corpses unburied after Philippine typhoon.* Retrieved December 30, 2013

[42] https://en.wikipedia.org/wiki/European_migrant_crisis, accessed Sept 2015.

[43] Justin Gillis, "Rising Sea Levels Seen as Threat to Coastal U.S.", NY Times, Mar 13, 2012.

[44] Dean Roemmich, "Argo and Ocean Heat Content: Progress and Issues", Scripps Institute of Oceanography, October 2013, CERES Meeting, Interestingly, the Southern oceans are warming much more than the Northern oceans, as one might expect with such a large ice cube (i.e. the Arctic ice) floating (and melting) in the Northern hemisphere.

[45] Vegar Schwartz, "Greenland ice mass balance using GRACE gravity data", Master's thesis, Norwegian University of Science and Technology, p 54; see also table 4.1 showing many other author's results.

[46] Tedesco et al, "Greenland Ice Sheet", Arctic Report Card: Update for 2014, NOAA, January 27, 2015.

[47] James Hansen et al, "Ice melt, sea level rise and superstorms: evidence from paleoclimate data, climate modeling, and modern observations that 2C global warming is highly dangerous", Atmos. Chem. Phys. Discuss., V15, #20059-201790, 2015.

[48] Glenn Scherer, "Climate Science Predictions Prove Too Conservative", Scientific American, Dec 2012.

[49] Hargreaves J. C., Annan J. D.. Can we trust climate models?. WIREs Clim Change2014, 5: 435-440. doi: 10.1002/wcc.288; and P Gosselin, "Last Ditch Clinging Effort... Scientists Plainly Have Struck Out on Short and Mid-Term Climate Model Reliability", Jan 2015; and AR5, IPCC, Chapter 9, "Evaluation of Climate Models".

[50] Looking at the IPCC reports, I speculate that one potential negative forcing that appears to have been left out (or is buried in various studies) is the energetic UV sunlight that goes through the Ozone hole(s), bounces off snow/ice, and then is reflected back out into

space, providing a cooling effect relative to where there is no ozone hole. This could potentially provide another explanation of why the Antarctic is colder than the Artic, when both polar regions should be more affected by global warming (as the Antarctic Ozone hole is larger than the Arctic's, it may provide more cooling.) *"A forcing at high latitudes yields a larger response than a forcing at low latitudes. This is expected because of the sea ice feedback at high latitudes and the more stable lapse rate at high latitudes"* (Hansen et al, "Radiative Forcing and Climate Response", J. Geophysical Research, Vol 102, Issue D6, 27 Mar 1997, pp 6831-6864.)

[51] http://www.skepticalscience.com/Climate-Change-The-40-Year-Delay-Between-Cause-and-Effect.html

[52] A good example of latency is how ice melts, but one example of phase changes that complicates climate modeling. The latent heat of ice is the extra energy needed to turn it from ice into liquid; during the time the ice is melting the temperature will stay constant. In fact, the latent heat of the glaciers' melting could even be considered a negative forcing that may have been neglected by climate models, and could therefore potentially be a simple explanation for the current "hiatus". World Glacier Monitoring Service Data (https://en.wikipedia.org/wiki/Glacier_mass_balance#/media/File:Bams_2013.jpg) shows a marked increase in glacier melt-off just as the "hiatus" began, circa 1998-2000. The higher level of glacial melt-off may have absorbed some of "missing heat".

[53] By Robert A. Rohde, Global Warming art project, https://commons.wikimedia.org/wiki/File:Holocene_Sea_Level. png , using data from: Fleming, Kevin, Paul Johnston, Dan Zwartz, Yusuke Yokoyama, Kurt Lambeck and John Chappell (1998). "Refining the eustatic sea-level curve since the Last Glacial Maximum using far- and intermediate-field sites". Earth and Planetary Science Letters 163 (1-4): 327-342; Fleming, Kevin Michael (2000). Glacial Rebound and Sea-level Change Constraints on the Greenland Ice Sheet. Australian National University. PhD Thesis; Milne, Glenn A., Antony J. Long and Sophie E. Bassett (2005). "Modelling Holocene relative sea-level observations from the Caribbean and South America". Quaternary Science Reviews 24 (10-11): 1183-1202.

[54] http://nca2014.globalchange.gov/

[55] http://www.arctic.noaa.gov/detect/ice-seaice.shtml

[56] http://news.bbc.co.uk/2/hi/europe/8264345.stm

[57] http://nrmsc.usgs.gov/research/glacier_retreat.htm

[58] The purple line is the 20-year average. National snow and ice data center, Boulder, CO.   http://www.arctic.noaa.gov/reportcard/sea_ice.html

[59] "Snowmelt Has Been Coming Earlier in Wyoming", earthobservatory.nasa.gov, Mar 2015.

[60] http://tidesandcurrents.noaa.gov/sltrends/sltrends.shtml

[61] http://www.pmel.noaa.gov/co2/story/Ocean+Acidification

[62] http://www.ncbi.nlm.nih.gov/pmc/articles/PMC2575286/

[63] http://www.cnn.com/2014/01/23/us/wacky-weather/index.html

[64] Derived from final figure, Jan Hollan, "No soon Ice Age, says astronomy", Dec 18, 2000.

[65] See http://biocycle.atmos.colostate.edu/shiny/Milankovitch/ for all graphs of all three types of forcings.

[66] Please see www.skepticalscience.com for a list of common Climate Myths and scientific evidence both for and against. Also, see https://www.aip.org/history/climate/cycles.htm for the best history of the global warming theory and how it developed over time

[67] U.S. Energy Information Administration

[68] U.S. Bureau of Economic Analysis, Real change in GDP chained 2009$ in blue; oil prices, plotted in red, from the U.S. Energy Information Administration

[69] U.S. Energy Information Administration, Annual and Quarterly Coal Reports.

[70] U.S. Energy Information Administration

[71] Energy Information Administration, historical coal prices and 2012/2013/2014 Annual Coal Reports.

[72] U.S. Energy Information Administration, Annual Coal Reports

[73] David Roberts, "Big Coal in big trouble as coal production costs rise", http://grist.org/climate-energy/big-coal-in-big-trouble-as-coal-production-costs-rise/ accessed 09/21/14.

[74] One example article: "Alpha joins the lineup of coal miners in bankruptcy", Aug 3, 2015, Associated Press.

[75] Kevin Jianjun Tu and Sabine Johnson-Reiser, "Understanding China's Rising Coal Imports", 2/16/2012, Carnegie Endowment for International Peace Policy Outlook

[76] FOB Newcastle/Port Kembia. www.indexmundi.com, accesses 09/21/14. Most Australian coal is exported to China.

[77] This is higher than gasoline prices increases due to economic ripple effects (the gasoline for all goods transportation costs more, etc.)

[78] U.S. Bureau of Economic Analysis. (Oil & Gas Extraction + Petroleum & Coal products Gross Output)/Total Gross Output.

[79] Kumudu Gunasekera & Brad Ship, "Trends in Highway and Rail Transit Construction Costs", June 2013, http://efr.pbworld.com/publications/default.aspx?id=24, accessed 9/15/2014.

[80] http://www.fao.org/worldfoodsituation/foodpricesindex/en/ accessed 9/24/14.

[81] Sharon O'Malley, "Running on empty: Federal road construction fund set to expire", May 2015, Construction Dive.

[82] As of 2013, per Gary Stoller, "U.S. roads, bridges are decaying despite stimulus influx", USA Today, Jul 29, 2013. There has been little movement since.

[83] Interestingly, the U.S. official inflation rate measure, the consumer price index, is heavily weighted toward food; so it is difficult to separate inflation from food prices.

[84] Amory B. Lovins and the Rocky Mountain Institute, "Reinventing Fire: Bold Business Solutions for the New Energy Era", 2011, Chelsea Green Publishing, White River Junction, VT, Chapter 1 "The True Cost of Oil Addiction."

[85] EIA. 18.57 Quad imported, average $102/barrel, 5.8x10^6 BTU/barrel, for 2012.

[86] See "Financial Cost of the Iraq War" on Wikipedia.org. These figures are likely conservative, and highly controversial, depending on whether indirect borrowing costs, veteran disability claims, equipment losses, etc.

[87] http://www.forbes.com/sites/afontevecchia/2013/02/05/bp-fighting-a-two-front-war-as-macondo-continues-to-bite-and-production-drops/

[88] Adding up the cost of oil and coal price increases on roads, food, electricity, extraction costs, imported oil cost, defense/cleanup, and war costs. This is admittedly imprecise.

[89] http://www.cbsnews.com/news/salt-water-fish-extinction-seen-by-2048/; for more recent events, see Aaron Guzman, "Massive Ocean Die-Offs in Pacific Northwest", www.fishfaqs.net

90 "Climate Change Evidence & Causes", National Academy of Sciences

91 See "The National Security Implications of a Changing Climate, May 2015, The White House, which merges findings from a number of DOD reports.

92 Henry Fountain, "Researchers Link Syrian Conflict to a Drought Made Worse by Climate Change", Mar 2, 2015, New York Times.

93 Per capita data taken from previous figure.

94 Bryan Walsh, "Why We Don't Care About Saving Our Grandchildren From Climate Change", Time, Oct 21, 2013.

95 Rhee, Nari. "The Retirement Savings Crisis: Is It Worse Than We Think?" National Institute on Retirement Security, June 2013.

96 Matos, G.R., 2012, Use of raw materials in the United States from 1900 through 2010: U.S. Geological Survey Fact Sheet 2012-3140, 7 p., available at http://pubs.usgs.gov/fs/2012/3140. (Supersedes Fact Sheet 2009–3008.) Data accessed 08/11/2015.

97 210 million gallons (Wikipedia.org: Deep Horizon) * 7.21 lbs/gallon * 1/2.205 kg/lbs *1 tonne/1000 kg ≈ 690,000 metric tons

98 Arthur Waskow, "Rabbis Against Climate Change", June 6, 2015, www.forward.com

99 http://islamicclimatedeclaration.org/islamic-declaration-on-global-climate-change/

100 Leviticus 25:23-24; Genesis 2:15; Genesis 1:26-28; and www.creationcare.org

101 Robert Fulghum, "All I Really Need to Know I Learned in Kindergarten: Uncommon Thoughts on Common Things", 1988, Ballantine Books.

102 Recent articles about total oil and gas reserves doubling by 2050 are also correct-- but the reserves referred to are "expensive to extract" oil shale and tar sands reserves. "Peak" oil is not about running out of oil, it is about running out of cheap oil.

103 Matt Smith, "The Mayor of Beijing Says His City Is 'Unlivable'", Jan 28, 2015, Vice news.

104 Dominique Patton, "More than 40 percent of China's arable land degraded: Xinhua", Nov 4th, 2014, Beijing, www.reuters.com

105 Dr. David Suzuki, "Toxic Smog Puts Cancer as Leading Cause of Death in China", April 15, 2015, www.ecowatch.com

106 Morgan Winsor, "China's Pollution Crisis: Nearly Two-Thirds of Underground Water is Graded Unfit For Human Contact, Report

Says", ibtimes.com, June 2015, citing a report by the Chinese Ministry of Environmental Protection.

[107] Stephen Chen, "Beijing drinking water reservoir had lead levels '20 times WHO standard' for at least three years", July 2015, South China Morning Post International Edition.

[108] Jonathan Kaiman, "China faces $176bn bill to clean up air pollution", Dec 20 2013, the Guardian; John Daly, "Bill for Cleaning China's Air Pollution - $290 Billion",

[109] David Stanway, "After China's multibillion-dollar cleanup, water still unfit to drink", reuters, 2/20/13

[110] "The Economics of China's Pollution Problem", May 1, 2013, Wharton Public Policy, University of Pennsylvania.

[111] http://www.tradingeconomics.com/china/gdp

[112] "2013 Draft to Congress on the Benefits and Costs of Federal Regulations and Agency Compliance with the Unfunded Mandates Reform Act", Office of Management and Budget (OMB), 2013. Note, this cost : benefit ratio can vary significantly, as the recent mercury regulation rejected by the Supreme Court showed.

[113] The low is from ibid; Greenstone et al, "The Effects of Environmental Regulation on the Competitiveness of U.S. Manufacturing", Sep 2012, MIT Center for Energy and Environmental Policy Research

[114] https://www.census.gov/foreign-trade/statistics/highlights/top/top1412yr.html and http://www.statista.com/statistics/263661/export-of-goods-from-china/. 467 $billion/2,342 $billion = 20%.

[115] Eurostat trade statistics.

[116] China's total consumption is ~ 36% of its GDP, or ~3.5 $ trillion in 2014 (worldbank.org). US exports are .5 trillion over the same period. .5 / (3.5 china consumption + 2.4 exports) = 8%, the U.S. consumer proportion of the total estimated $500 billion annual cleanup costs. This is ~40 $billion dollars annually.

[117] Craig Simons, "China's Rise Creates Clouds of U.S. Pollution", 2011, http://aliciapatterson.org; and "Chinese Air Pollution Reaches US West Coast", Jeff, Jan 25, 2013, http://www.8asians.com/2013/01/25/chinese-air-pollution-reaches-us-west-coast/

[118] The higher 2% figure from Clyde Crews Jr. "Tip of the Costberg", Competitive Enterprise Institute, 2016; the low from OMB estimates of regulatory costs + EPA's annual budget of ~8 $billion.

There are references that allude to higher costs (for example, Greenstone et al "The Effects of Environmental Regulations on the Competitiveness of U.S. Manufacturing", Sep 2012). But what these analyses usually fail to take into account is the ability of innovation to reduce compliance costs.

[119] Ron Bousso and Karolin Schaps, "Update 2-Oil bosses to meet in latest climate change offensive", Oct 7, 2015, reuters.com

[120] William Nordhaus, "The Climate Casino: Risk, Uncertainty, and Economics for a Warming World", Yale University Press, 2013, Chapter 25, pp 278.

[121] ibid, Chapter 20.

[122] See worldwide emissions in Figure 23

[123] See "European Union Emissions Trading Scheme", www.wikipedia.org, accessed 10/8/14

[124] See also William Nordhaus, "The Climate Casino: Risk, Uncertainty, and Economics for a Warming World", Yale University Press, 2013, Chapter 21 for a detailed comparison of the pros, cons, and differences between these two policies, and why he believes a carbon tax is preferable.

[125] See Alan Durning and Yoram Bauman, "All You Need to Know About BC's Carbon Tax Shift in Five Charts", www.sightline.org; www.wikipedia.org "British Columbia carbon tax"; and "British Columbia's Carbon Tax Shift: The First Four Years", Sustainable Prosperity, June 2012.

[126] Marlo Lewis, "Why British Columbia's Carbon Tax is Not Applicable to America", Competitive Enterprise Institute, 09/16/2014

[127] U.S. Energy Information Administration

[128] U.S. Federal Highway Administration and Census Bureau

[129] Except opportunity cost.

[130] Stats.oecd.org, constant dollars and purchasing power index (PPP). Note, population trends, especially in Japan, have also impacted economic growth over this time period.

[131] U.S. Federal Register, Vol 80, No 205, Oct 23, 2015, p 64532, "C. Affected Units"

[132] See Peter Van Doren's "A Brief History of Energy Regulations", February 2009, the Cato Institute, www.downsizing government.org/energy/regulations for further historical energy regulation distortions

[133] See https://www.fhwa.dot.gov/infrastructure/gastax.cfm for a history of the gas tax in the U.S.

[134] "Increasing the Efficiency of Existing Coal Fired Power Plants", Richard Campbell, 12/20/2013, Congressional Research Service, R43343, p 21.

[135] Riccardo Ambrosini, "Profiles: Life extension of coal-fired power plants", IEA Clean Coal Centre, December 2005.

[136] "Ripe for Retirement: The Case for Closing America's Costliest Coal Plants", Union of Concerned Scientists, Dec 2013.

[137] EIA. Data from "Age of electric power generators varies widely", June 16, 2011. Pre 2010, from Form EIA-860 Annual Electric Generator Report, and Form EIA-860M (see Table ES3 in March 2011 Electric Power Monthly).

[138] Actually, plants could not burn just natural gas or petroleum, they had to have the capability to burn coal or alternatives, too. But this made natural gas plants too expensive, essentially banning their construction.

[139] Points 1 thru 4 are cited from: William Nordhaus, "The Climate Casino: Risk, Uncertainty, and Economics for a Warming World", Yale University Press, 2013, Chapter 22.

[140] See the Energy Policy Act of 1992, and further extensions of the PTC (Production Tax Credit).

[141] Alan Goodrich, Ted James, and Michael Woodhouse, "Solar PV Manufacturing Cost Analysis: U.S. Competitiveness in a Global Industry", October 10, 2011, National Renewable Energy Lab (NREL).

[142] Eric Wesoff, "SolarWorld Wins Again: Big Anti-Dumping Tariffs in US-China Solar Panel Trade Case", www.greentechmedia, July 25, 2014.

[143] Shelagh Whitley, "Time to Change the Game: Fossil Fuel Subsidies and Climate", Overseas Development Institute, Nov 2013; "Taxing Energy Use, A Graphical Analysis", OECD, 2013; IEA World Energy Outlook 2012.

[144] Makhijani et al, "Cashing in on "All of the Above": U.S. Fossil Fuel Production Subsidies Under Obama", Oilchange International, July 2014.

[145] USA FFSR Progress Report to G20 2014 Final, U.S. Treasury Department, "United States – Progress Report on Fossil Fuel Subsidies"

[146] Brad Plumer, "The U.S. will stop financing coal plants abroad. That's a huge shift." The Washington Post, June 27th, 2013.

[147] http://www.heritage.org/research/reports/2013/05/a-farm-bill-primer-10-things-you-should-know-about-the-farm-bill; and http://www.ewg.org/downfall-direct-payments

[148] Congressional Budget Office (CBO), "Federal Financial Support for the Development and Production of Fuels and Energy Technologies", Mar 2012; Adeyeye et al, "Estimating U.S. Government Subsidies to Energy Sources: 2002-2008", Environmental Law Institute, Sep 2009; and "60 Years of Energy Incentives: Analysis of Federal Expenditures for Energy Development", Management Information Services, Inc., Oct 2011, for The Nuclear Energy Institute

[149] Tim Worstall, "Renewables Get 25 Times the Subsidy That Fossil Fuels Do", 11/13/2013, www.forbes.com

[150] Nancy Pfund and Ben Healey, "What would Jefferson do? The Historical Role of Federal Subsidies in Shaping America's Energy Future", DBL Investors, Sep 2011.

[151] "60 Years of Energy Incentives: Analysis of Federal Expenditures for Energy Development", Management Information Services, Inc., Oct 2011, for The Nuclear Energy Institute

[152] The Energy Tax Act of 1978

[153] Called the Volumetric Ethanol Excise Tax Credit (VEETC)

[154] "Corn Ethanol Subsidies Are Alive and Well", Oct 16, 2013, Taxpayers for CommonSense.

[155] James Conca, "It's Final – Corn Ethanol Is Of No Use", Forbes, April 20, 2014; and Charles et al, "Biofuels – At What Cost? A review of costs and benefits of EU biofuel policies", Institute for Sustainable Development, April 2013.

[156] Dr. Thomas E. Elam, FarmEcon LLC, "Food Costs are Eating American Family Budgets", Jan 2013.

[157] James Acton and Mark Hibbs, "Why Fukushima was Preventable", March 2012, The Carnegie Papers.

[158] Tracy Loew, "Fukushima radiation has reached North American shores", 4/6/2015, Statesman Journal. See also www.netc.com

[159] http://en.wikipedia.org/wiki/Nuclear_reactor_accidents_in_the_United_States

[160] Summer 2 and 3 in South Carolina, and Vogtle 3 and 4, in Georgia.

[161] Kenichi Oshima, "Fukushima disaster bill more than $105bn, double earlier estimate – study", www.rt.com, Aug 2014,

[162] William Nordhaus, "The Climate Casino: Risk, Uncertainty, and Economics for a Warming World", Yale University Press, 2013, Chapter 22.

[163] See also David Austin "Addressing Market Barriers to Energy Efficiency in Buildings", Congressional Budget Office, August 2012

[164] Fossil fuel extraction costs (See Figure 15) are over 4X more than the <1% GDP needed to solve global warming.

[165] There are, of course, population dynamics and other forces that impact the lower growth of the EU and Japan; but higher fossil fuel costs are one key factor influencing growth per White's law.

[166] The CBO is right that only a tiny fraction of corn prices finds its way into food prices – much more is spent on transportation, advertising, packaging, etc. However, use of ethanol also directly impacts gasoline prices (both directly, and by virtue that the energy content in ethanol is lower requiring more fill-ups per year to go the same distance), as well as restaurant food prices (~1/3 of U.S. food budgets). See "The Renewable Fuel Standard: Issues for 2014 and Beyond", CBO, June 2014; "Federal Ethanol Policies and Chain Restaurant Food Costs", PwC, Nov 2012; USDA estimates of U.S. food expenditures; Thomas Landstreet, "High Food Prices: An Investor's Dilemma", Forbes, June 2014.

[167] Emissions data: Carbon Dioxide Information Analysis Center (CDIAC), www.cdiac.ornl.gov. Atmospheric emissions are Mauna Loa(1950+), and Law Dome ice cores prior to 1950. 2007+ population: www.prb.org; prior to 2007, www.scottmanning.com, averages of multiple population estimates. UN projections: www.esa.un.org, world population prospects.

[168] Matthew Connelly, "Fatal Misconception: The Struggle to Control World Population", Belknap Press of Harvard University Press, Cambridge, MA, 2008.

[169] "Population Control: Real Costs, Illusory Benefits", by Stephen Mosher, Transaction Publishers 2008, is another good resource, written from a more conservative viewpoint.

[170] With some obvious exceptions. In 1939, 17 Million Jews were reduced to 11 million in the Holocaust, and their population has only partially recovered (13.3 million in 2013). (see

http://www.simpletoremember.com/vitals/world-jewish-population.htm for further information.)

[171] U.S. Census, average of foreign born population over last twenty years (2010 and 1990).

[172] As a side note, Limits to Growth contributed to China's one child policy: see Matt Ridley, "China's one-child policy was inspired by western greens", Jan 18, 2014, www.rationaloptimist.com; and Martin King Whyte, Wang Feng, and Yong Cai "Challenging Myths About China's One-Child Policy", The China Journal, #74, 1324-9347, 2015, Australian National University.

[173] see http://www.independent.co.uk/news/world/middle-east/israel-gave-birth-control-to-ethiopian-jews-without-their-consent-8468800.html and
http://www.haaretz.com/news/national/israel-admits-ethiopian-women-were-given-birth-control-shots.premium-1.496519#!

[174] Quote from Imam Hossain, in "Resolution Re Steps to Check the Increase in Population", Mar 18 1935, Council of State Debates: Official Report, 1936, vol 1, #15, IOR, V/9/244. {Reference from Chapter 3, Matthew Connelly, "Fatal Misconception: The Struggle to Control World Population", Belknap Press of Harvard University Press, Cambridge, MA, 2010.}

[175] www.usaid.gov/results-and-data, accessed 11/3/14

[176] Matthew Connelly, "Fatal Misconception: The Struggle to Control World Population", Belknap Press of Harvard University Press, Cambridge, MA, 2008, pp 20-35; 90-110, and 340-350.

[177] United Nations Department of Economic and Social Affairs Population Division, Fertility and Family Planning Section, World Fertility Data 2013

[178] United Nations Department of Economic and Social Affairs Population Division, Fertility and Family Planning Section, Statistical Annex Table 3.A, 2013

[179] United Nations Population Division

[180] United Nations Population Division

[181] UN Population Division, 2013 projections. The 2052 Update to Limits to Growth forecast data is found at www.2052.info, and comes from Jorgen Randers, "2052: A Global Forecast for the Next Forty Years, A Report to the Club of Rome Commemorating the 40th Anniversary of The Limits to Growth," Chelsea Green Publishing, White River Junction, VT, 2012.

[182] "Industry Wide Trends for Cost Reduction of Crystalline Silicon Modules", Graham Stevens, Intersolar North America, San Francisco, CA, July 2008.

[183] http://www.ieeeghn.org/wiki/index.php/Pearl_Street_Station

[184] http://news.xinhuanet.com/fortune/2013-09/25/c_117508544.htm, which describes one example of a high capacity transmission line in China

[185] "Innovation in Wind Turbine Design" by Peter Jamieson, 2011, John Wiley & Sons, Ltd, Table 1.2 for these ratios.

[186] See www.dfma.com

[187] Inventory of U.S. Greenhouse Gas Emissions and Sinks, 1990-2010. April 15, 2010; EPA 430-R-12-001.

[188] http://www.epa.gov/cleanenergy/energy-and-you/affect/air-emissions.html; also, see "Cost and Performance Baseline for Fossil Energy Plants, Volume 1: Bituminous Coal and Natural Gas to Electricity", rev 2a, Sep 2013, DOE/NETL-2010/1397, U.S. DOE. Comparing case 1 and 13 on a $/MWh(net), p5, yields $804/1723 = 46.6\%$ lower CO_2 emissions for a baseline Natural Gas plant

[189] Energy Information Administration (EIA)

[190] www.wikipedia.org/wiki/Engine_efficiency; www.wikipedia.org/wiki/Combined_cycle; www.wikipedia.org/wiki/Cogeneration; accessed September 2015.

[191] www.wikipedia.org, terms: Otto_engine, four-stroke engine, fuel efficiency, Ford EcoBoost engine, diesel engines, accessed September 2015; and Takaishi et al "Approach to High Efficiency Diesel and Gas Engines", Mitsubishi Heavy Industries Ltd, Technical Review Vol 45, #1, Mar 2008. Fuel efficiency data from U.S. DOT, Model T specifications, antique automobile club of America, and www.anythingaboutcars.com. Specific data for 1920-1970 is relatively unreliable.

[192] See ""The Opportunity for CHP in the United States", May 2013, American Gas Association.

[193] Note, fuel cell cars have the potential to reduce CO_2, but may not in reality have this property, when one considers all the myriad systems needed to run a fuel cell car (i.e. natural gas reforming to obtain H_2, etc.)

[194] When natural gas pipelines are not readily accessible to a drilling site, the industry flares the gas instead (properly converting it to CO_2 rather than allowing 20X more powerful methane to be released), because the cost of installing a pipeline is much higher than the value of the gas flared.

[195] Note, competitiveness is strongly affected by renewables and fossil fuel subsidies, the absence of a price on carbon, etc.

[196] http://rredc.nrel.gov/wind/pubs/atlas/maps.html

[197] Denmark, Germany, and California are leaders in demonstrating the technical practicality of this level of renewables. Utilities in the U.S. are constantly balancing fluctuating loads with more constant base-load electricity supplies. Introducing fluctuating sources (with the sun and wind) introduces more complexity with regard to this balancing act, but is eminently feasible, as many utilities in the U.S. have demonstrated.

[198] http://en.wikipedia.org/wiki/Nuclear_reactor_accidents_in_the_ United_States

[199] http://en.wikipedia.org/wiki/Nuclear_power_in_the_United _States

[200] U.S. Energy Information Administration. Forecasts are from the Reference Case, AEO 2014; historical from the Short Term Energy Outlook, June 2014, adjusted into 2012 $; and the AEO 2015 Reference Case (April 2015).

[201] "Increasing Distillate Production at U.S. Refineries – Past Changes and Future Potential", US EIA, Oct 2010

[202] For example, gasoline stores 44.1 MJ/kg; 32.7 MJ/Liter[202] , and a large scale hydro plant (~300 feet high) stores 1 MJ/kg; .001 MJ/Liter[202]. A single gallon of gasoline contains more stored energy than 32,000 gallons of water 300 feet above you.

[203] At an estimated 200 $/kWh, the 85 kWh Tesla 250 mile range battery (it can go 300 miles on one charge, but this kills battery life), with an estimated 6 year lifetime for 80% deep discharge Li-ion batteries, would cost $17,000. At U.S. average .11 $/kWh electricity cost, this means spending $2833 (battery) + $476 (electricity) annually to drive an average 15000 miles, or .22 $/mile. This is 45% higher than a 20 mpg vehicle @ 3 $/gallon.

[204] See Amory Lovins, "Winning the Oil Endgame", Sep 2004, Rocky Mountain Institute.

[205] "Light-Duty Automotive Technology, Carbon Dioxide Emissions, and Fuel Economy Trends: 1975 Through 2013", Executive Summary, EPA, Dec 2013.

[206] Glenn Meyers, "Hydrogen Economy: Boom or Bust?", Mar 19th, 2015, www.cleantechnica.com

[207] "Wood Based Energy and Carbon Neutrality in Oregon", Oregon Forest Biomass Working Group, 2011

[208] http://www.afdc.energy.gov/fuels/prices.html

[209] USDA, "Measuring the Indirect Land-Use Change Associated With Increased Biofuel Feedstock Production, A Review of Modeling Efforts, Report to Congress", Feb 2011

[210] See Nicolas Loris, "The Ethanol Mandate: Don't Mend It, End It", Jun 2013, The Heritage Foundation; Christopher Knittel, "Corn Belt Moonshine The Costs and Benefits of U.S. Ethanol Subsidies", American Boondoggle, July 2011.

[211] William Nordhaus, "The Climate Casino: Risk, Uncertainty, and Economics for a Warming World", Yale University Press, 2013, Chapter 22, for further discussion. Nordhaus cites studies that CAFÉ standards cost much more than carbon taxes with regard to reducing emissions; however, this view is wrong because there is so much low hanging fruit (which exists because carbon prices are currently zero). Real answers hinge on engineering cost vs. efficiency curves, which inevitably turn out to be less steep than initially projected due to lean engineering; and projected future gas costs and discount rate assumptions. Relatively low engineering cost vs. efficiency curves, 4% discount rates, and projected higher gasoline costs point toward accelerating mpg standards.

[212] There are some processes that reconvert the absorption products back to absorbent + sulfur or gypsum; these byproducts are then sold to offset the costs of capture. Currently these regenerating processes are <10% of the market.

[213] "Economic and energetic analysis of capturing CO_2 from ambient air", House, Baclig, Ranjan, Nierop, Wilcox, & Herzog, pnas.org, September 14, 2011. Note, the assumptions in this study are pessimistic. They assume 5% actual:theoretical efficiency, and .010 $/kWh for electricity. Graphs from the same study show some 10% processes, and industrial electricity without transmission costs

is likely to cost ~.05 $/kWh. This equates to ~125 $/tonne as shown in Figure 38.

[214] Concentration: Average of "Abundance in Earth's Crust", WebElements.com; "List of Periodic Table Elements Sorted by Abundance in Earth's Crust", http://www.science.co.il/PTelements.asp?s=Earth; and Jefferson Lab "It's Elemental" Periodic Table of the Elements. Costs: www.metalprices.com, 2010-2015 averages. CO_2: Jeremy David & Howard Herzog, "The Cost of Carbon Capture", MIT, 2000 and "Economic and energetic analysis of capturing CO_2 from ambient air", House, Baclig, Ranjan, Nierop, Wilcox, & Herzog, pnas.org, September 14, 2011. SO_x: Paul Nolan, "Flue Gas Desulfurization Technologies for Coal-Fired Power Plants", Nov 2000, Babcock & Wilcox, Coal-Tech 2000 International Conference. NO_x: "Cost estimate of deNOx for a flue gas from MAN B&W marine engine", Technical University of Denmark, Risø 2011.

[215] http://www.canadiancleanpowercoalition.com/pdf/CO2%20Transportation%20Cost%20Calculations.pdf, June 2012. Sources include NETL, and the Oil and Gas Journal.

[216] Figure 3 in Interstate Oil and Gas Compact Commission, Bliss et al, "A Policy, Legal, and Regulatory Evaluation of the Feasibility of a National Pipeline Infrastructure for the Transport and Storage of Carbon Dioxide", Sep 10, 2010. Figure by Steve Melzer, Melzer Consulting, used with permission. See www.melzerconsulting.com for further information regarding CO_2 pipelines in the U.S.

[217] "Carbon Dioxide Enhanced Oil Recovery: A Critical Domestic Energy, Economic, and Environmental Opportunity", National Enhanced Oil Recovery Initiative, Center for Climate and Energy Solutions, Feb 2012, p. 1.

[218] See http://pubs.usgs.gov/dds/dds-81/Intro/facts-sheet/GasKillingTrees.html, which describes tree die-off on Mammoth Mountain, CA due to CO_2 seepage;and Atlas Obscura, "Lake Nyos Suffocated Over 1,746 People in a Single Night", www.slate.com. Lake Nyos is a crater lake in Cameroon, and suddenly released high levels of CO_2 in August, 1986, suffocating people as far as 15 miles away.

[219] U.S. DOE. 2008. Carbon Cycling and Biosequestration: Integrating Biology and Climate Through Systems Science; Report from the March 2008 Workshop, DOE/SC-108, U.S. Department of Energy Office of Science (genomicscience.energy.gov/carboncycle/).

[220] Abel Mendez, "Distribution of landmasses of the Paleo-Earth", Jul 6, 2011, http://phl.upr.edu/library/notes/distribution oflandmassesofthepaleo-earth

[221] http://en.wikipedia.org/wiki/Keeling_Curve

[222] Source: Narayanese and Sémhur, Mauna Loa Hawaii Observatory, noaa.gov. http://en.wikipedia.org/wiki/File:Mauna_Loa_Carbon_Dioxide_Apr2013.svg

[223] Colin Austin, "Resolving Climate Change 3, How Science Can Fail Us", www.waterright.com.au, Sep 2012.

[224] Kulyk, Nataliya, "Cost-Benefit Analysis of the Biochar Application in the U.S. Cereal Crop Cultivation", 2012, Center for Public Policy Administration Capstones, Paper 12, University of MA.

[225] Shackley et al, "An Assessment of the Benefits and Issues Associated with the Application of Biochar to Soil". UK Biochar Research Centre, 2011

[226] Galinato et al, "The Economic Value of Biochar in Crop Production and Carbon Sequestration". Energy Policy 39 (2011).

[227] http://www.ars.usda.gov/is/AR/archive/feb01/bank0201.htm

[228] "Fertilizers and Climate Change: Enhancing Agricultural Productivity and Reducing Emissions", 24 July 2009, www.fertilizer.org (a worldwide fertilizer industry organization).

[229] http://www.nature.com/scitable/knowledge/library/soil-carbon-storage-84223790

[230] International Panel on Climate Change, AR5, Chapter 6-3, anthropogenic emissions between 1750 and 2011 were 545 +/- 85 PgC, and land use change (including deforestation, afforestation, and reforestation) contributed 180+/-80 PgC. 180/545 = 33%.

[231] "Carbon in the Vegetation and Soils of Great Britain", Milne and Brown, Journal of Environmental Management (1997) 49, 413–433

[232] "66 Ways to Absorb Carbon and Improve the Earth's Reflectivity, From Reasonable Options to Mad Scientist Solutions", Risto Isomaki, Into Publishing, 2009, Way #8.

[233] "66 Ways to Absorb Carbon and Improve the Earth's Reflectivity, From Reasonable Options to Mad Scientist Solutions", Risto Isomaki, Into Publishing, 2009, Way #8.

[234] http://www.unep-wcmc.org/natural-fix_61.html

[235] Report of the Interagency Task Force on Carbon Capture and Storage, Aug 2010, www.epa.gov; CCS Cost Workshop, 22-23 Mar 2011, MIT; Mike Fowler et al, "How Much Does CCS Really Cost?", The Clean Air Task Force, Boston, MA, Dec 2012.

[236] Quoted with permission. Found in both Suck It Up: "How capturing carbon from the air can help solve the climate crisis", Marc Gunther, 2012; and Google group comments on the report "Direct Air Capture of CO_2 with Chemicals, A Technology Assessment for the APS Panel on Public Affairs", June 1, 2011, APS Physics; (https://groups.google.com/forum/#!msg/ geoengineering/tyA68q_pcaY/umESINfSyZEJ)

[237] See, for example, NRDC's "Closing the Power Plant Carbon Pollution Loophole: Smart Ways the Clean Air Act Can Clean Up America's Biggest Climate Polluters", bracketing efficiency improvement scenarios between 5-10%.

[238] Amory B. Lovins and the Rocky Mountain Institute, "Reinventing Fire: Bold Business Solutions for the New Energy Era", 2011, Chelsea Green Publishing, White River Junction, VT, Chapter 1 "The True Cost of Oil Addiction."

[239] Williams, J.H., B. Haley, F. Kahrl, J. Moore, A.D. Jones, M.S. Torn, H. McJeon, "Pathways to deep decarbonization in the United States", Nov 2014, by Energy and Environmental Economics Inc. (E3), Lawrence Berkeley National Laboratory (LBNL), and Pacific Northwest National Laboratory (PNNL), for the Sustainable Development Solutions Network and the Institute for Sustainable Development and International Relations.

[240] MacDonald et al, "Low Cost and Low Carbon Wind and Solar Energy Systems: Feasible with Large Geographic Size", NOAA, Earth System Research Laboratory, 27 May 2014

[241] David MacKay, "Sustainable Energy – without the hot air", 2009, UIT, Cambridge, England. See www.withouthotair.com.

[242] See http://thesolutionsproject.org/

[243] Digging deeper into assumptions, the savings are dwarfed by additional transportation fuel costs (electrolyzer efficiency is assumed to be too high; electrolytic compressed hydrogen from

water actually costs a lot more than assumed; discount rates are too low; and wind capacity factors to generate H_2 are 2X-3X what they could be in reality); and thermal storage in soil or CSP wastes 10-50% of the energy ("Thermal Energy Storage Technology Brief", IRENA, Jan 2013), so these storage solutions are not practical. In addition, wind does not blow in the summer and in the southeastern U.S., so Jacobson's plan for the SouthEast at night does not appear credible. See also http://energyskeptic.com/2015/winds-dirty-secret-stops-working-across-continental-usa/

[244] See also Ulf Bossel & Baldur Eliasson, "Energy and the Hydrogen Economy", Jan 2003. It's a bit dated, but the energy considerations still apply.

[245] Phillip Brown & Gene Whitney, "U.S. Renewable Electricity Generation: Resources and Challenges", Congressional Research Service, August 2011; Carnegie et al, "Utility Scale Energy Storage Systems Benefits, Applications, and Technologies", June 2013, State Utility Forecasting Group. David Murphy and Charles Hall, "Year in Review—EROI or Energy Return on Energy Invested", Annals of the New York Academy of Sciences, Volume 1185, 2010, pp 102-118, New York Academy of Sciences; Weisbach et al, "Energy intensities, EROIs, and energy payback times of electricity generating plants", Energy, V52, Apr 2013, pp 210-221.

[246] John Upton, "More nukes: James Hansen leads call for 'safer nuclear' power to save climate", Nov 2013, grist.org

[247] James Acton and Mark Hibbs, "Why Fukushima was Preventable", March 2012, The Carnegie Papers.

[248] http://www.deepseanews.com/2013/11/true-facts-about-ocean-radiation-and-the-fukushima-disaster/

[249] Each new nuclear plant is a design unto itself, especially after Fukushima safety design changes. But the primary variable for product costing (including nuclear plants) is production volume. If nuclear plant construction costs are to be reduced, one needs to build a lot of plants with the exact same design to obtain economies of scale. France appears to have done this when it built its plants in the 1980s, but not since.

[250] http://www.wipp.energy.gov/WIPPRecovery/path_forward.html

[251] Solow-Swan model; endogenous growth theory by Paul Romer & Robert Lucas; Unified growth theory; and Schumpeterian growth

231

theory. See www.wikipedia.org, under "Economic Growth" for further reading.

[252] Leslie White, "The Evolution of Culture: The Development of Civilization to the Fall of Rome", 1959.

[253] See www.wikipedia.org, "List of recessions in the United States"

[254] "Denard scaling deals with switching speeds and other physical characteristics of transistors, and thus heat dissipation and maximum clock speeds", from the next endnote.

[255] Joel Hruska, 8/30/2013, "Intel's former chief architect: Moore's law will be dead within a decade", www.extremetech.com, accessed 8/26/2014.

[256] Sunil Kanwar and Robert Evenson, "Does intellectual property protection spur technological change?" Oxford Economic Papers, Vol 55, 2003, pp 235-264.

[257] Atta et al, "Commercial Industry Research & Development Management Best Practices", Institute for Defense Analysis, Dec 2011.

[258] See http://www.clustermapping.us/region; www.jimgollub.com; Michael Porter "Clusters and the New Economics of Competition", Harvard Business Review, Nov-Dec 1998; Richard Florida "Who's Your City?", Basic Books, 2008; "Cluster-Based Economic Development: A Key to Regional Competitiveness", Economic Development Administration, October 1997; and "Business cluster" on www.wikipedia.org

[259] In many developing countries, 3% of the population pay income taxes, as opposed to 60-80% for developed countries. The main source of tax revenue is export/import/customs, collected when goods pass through the border.

[260] Peter Lloyd, "Free Trade and Growth in the World Economy", University of Melbourne; and James K. Jackson, "Trade Agreements: Impacts on the U.S. Economy", April 10, 2013, Congressional Research Service

[261] See Sarah Chayes, "Thieves of State: Why Corruption Threatens Global Security, WW Norton, New York, 2015; and Zephyr Teachout, "Corruption in America: From Benjamin Franklin's Snuff Box to Citizens United", Harvard University Press, Cambridge, MA, 2014

[262] For more information, see the U.S. Bureau of Economic Analysis, www.bea.gov. See also Mark Skousen, "At Last, a Better Economic Measure", April 22, 2014, the Wall Street Journal.

[263] U.S. Bureau of Economic Analysis. Resource Extraction is the sum of Mining, Oil &Gas, Support activities for mining, and Petroleum and Coal Product categories; Computer Revolution is the sum of Data processing, warehousing and storage (aka Amazon), and computer systems design; Health care is the sum of Hospitals, Ambulatory Health Care Services, Nursing, and Insurance Carriers & related activities.

[264] Recognizing that Gross Output is a blunt measurement; and that this increase includes all mining, not just coal.

[265] U.S. Bureau of Economic Analysis

[266] "2014 U.S. Approval of Congress Remains Near All-Time Low", Rebecca Riffkin, Dec 15, 2014, Gallup U.S. Daily, www.gallup.com

[267] U.S. Government Accountability Office, per http://www.dodig.mil/resources/fraud/fraud_defined.html, accessed 2/26/15

[268] U Myint, "Corruption: Causes, Consequences, and Cures", Asia-Pacific Development Journal, Vol 7, #2, Dec 2000; Section I: Definition and Concepts.

[269] See Industry Week, "Can Lean Six Sigma Reduce Government Waste?" edited by Jill Jusko, Sep 20, 2011.

[270] 1 Timothy 6:10, King James version.

[271] Letter to Bishop Mandell Creighton, April 5, 1887 published in "Historical Essays and Studies", edited by J. N. Figgis and R. V. Laurence; London, Macmillan, 1907.

[272] See, for example: (1) Henry Bourne, "Food Control and Price-Fixing in Revolutionary France", the Journal of Political Economy, V27, #2, Feb 1919. (2) "25 Years of Transition: Post-Communist Europe and the IMF", James Roaf et al, the International Monetary Fund, Oct 2014. (3) "The Shadow Economy: An International Survey", Friedrich Schneider, Dominik Enste, Cambridge University Press, Feb 2013.

[273] U Myint, "Corruption: Causes, Consequences, and Cures", Asia-Pacific Development Journal, Vol 7, #2, Dec 2000, p 52. "Low income countries usually have highly regulated economies that give rise to large monopoly rents. Accountability in these countries is generally weak. Political competition and civil liberties is often

restricted. Laws and legal systems are generally weak. Watchdog agencies/the press are not well developed and sometimes suppressed. Discretionary powers of administrators are large...." All of these systemic conditions point to higher levels of corruption.

[274] Kirk Nicholas, Director of CPI/LSS programs, U.S. Army – Office of Business Transformation, http://www.idga.org/intelligence/webinars/army-transformation-deploying-lean-six-sigma/; see also "Lean Six Sigma Is in the Army Now, Improving Efficiency", Elaine Schmidt, www.isixsigma.com

[275] Kellen Giuda, "Lean Government Six Sigma? Why Do Politicians Ignore It?", 9/09/2012, Forbes.

[276] "2014 Annual Report on Business Transformation Providing Readiness at Best Value", 13 January 2014, Department of the Army.

[277] https://www.federalregister.gov/uploads/2015/05/Federal-Register-Pages-Published-1936-2014.pdf

[278] AFLCIO. www.aflcio.org/Corporate -Watch/Paywatch-2014, analysis of 350 companies in the S&P 500.

[279] George Reisman, "Piketty's Capital: Wrong Theory Destructive Program", TJS Books, Laguna Hills, CA, 2014, p 43 (location 763 of 1186)

[280] Robert Frank, "A Remedy Worse than the Disease: Why Higher Taxes are Better than Pay Caps", Stanford Center for Poverty and Inequality, Summer 2010, p 18.

[281] Xavier Gabaix & Augustin Landier, "Why Has CEO Pay Increased So Much?", 2008, Quarterly Journal of Economics, V123, Issue #1, pp 49-100, 2008 by the President and Fellows of Harvard College and the Massachusetts Institute of Technology.

[282] Ibid.

[283] Michael Hiltzik, "The right way to measure CEO pay has nothing to do with 'shareholder value'", LA Times, May 4, 2015.

[284] Lucian Arye Bebchuk & Jesse M. Fried, "Executive Compensation as an Agency Problem", Journal of Economic Perspectives, V17, #3, Summer 2003, pages 71-92.

[285] Lucian Arye Bebchuk & Jesse M. Fried, "Pay without Performance, The Unfulfilled Promise of Executive Compensation", 2004, Chapter 4.

[286] Hugh Rockoff, "Price Controls", the Concise Encyclopedia of Economics, 2nd Ed., Dec, 2007, www.econlib.org/library/Enc/PriceControls.html

[287] Section 162(m) of the U.S. tax code imposes limits on tax deductions for normal pay over $1million, but does not impose limits on "performance-based" pay, such as stock options. Companies switched to paying their executives with more stock options to avoid this limit.

[288] Lucian Arye Bebchuk & Jesse M. Fried, "Executive Compensation as an Agency Problem", Journal of Economic Perspectives, V17, #3, Summer 2003, pages 71-92.

[289] Items V and VI are threshold value creation criteria to ensure that executives do not pillage our natural resources and conduct business ethically, rather than promoting these two causes.

[290] A good example of this relative to market capitalization is provided by Aswath Damodaran in his "Musing on Markets" blog "Is your CEO worth his (her) pay? The Pricing and Valuing of Top Managers!", Apr 29, 2015. http://aswathdamodaran.blogspot.com/2015/04/is-your-ceo-worth-his-her-pay-pricing.html. This type of analysis needs to be extended to categories other than (i).

[291] "How Do Patients Choose Physicians? Evidence from a National Survey of Enrollees in Employment-Related Health Plans", Katherine Harris, Health Services Research, 2003 April, Vol 38, #2, p 711-732.

[292] Medical graduate data from: Association of American Medical Colleges. Composite of data from "The Complexities of Physician Supply and Demand: Projections Through 2025: Center for Workforce Studies", Michael J. Dill & Edward S. Salsberg, Nov 2008 (pre-2006); and "Table 27: Total Graduates by U.S. Medical School, Sex, and Year" from www.aamc.org, 2005-2014, for years 2006+. Population data from the U.S. Census Bureau.

[293] Todd Hixon, "Why are US Health Care Costs So High?", Forbes, March 1, 2012

[294] Mossialos et al, editors, "2014 International Profiles of Health Care Systems", January 2015, The Commonwealth Fund pub. No. 1802.

[295] The College Board reports average public college costs of $23000/year, and private college costs $46,000/year. With 73% of students attending public colleges

(http://www.usnews.com/education/blogs/the-college-solution/2011/09/06/20-surprising-higher-education-facts), this implies ~ $30,000 annually, or $120,000 overall.

[296] The Association of American Medical Colleges estimates the costs of 2014 private medical school education at $298,538 and $226, 447 public, with a recent AAMC Graduation Questionnaire citing ~50% public, 50% private. This averages to roughly $260,000. Source: AAMC October 2014 Debt Fact Card, and "Physician Education Debt and the Cost to Attend Medical School 2012 Update", Feb 2013, www.aamc.org.

[297] Starfield et al, "The Effects of Specialist Supply on Populations' Health: Assessing the Evidence", http://content.healthaffairs.org/cgi/content/abstract/hlthaff.w5. 97v1

[298] Chris Conover, "Are U.S. Doctors Paid Too Much?" May 28, 2013, Forbes.

[299] The Balanced Budget Act of 1997 maintains Medicare funding for Graduate Medical Education at 1996 levels, thereby acting as a ceiling on physician supply.

[300] Mossialos et al, editors, "2014 International Profiles of Health Care Systems", January 2015, The Commonwealth Fund pub. No. 1802.

[301] "American Schools vs. the World: Expensive, Unequal, Bad at Math", Julie Ryan, The Atlantic, Dec 2013

[302] Jal Mehta, "The futures of school reform: Five pathways to fundamentally reshaping American schooling", American Enterprise Institute, Nov 14, 2012

[303] Valerie Strauss, "Are private schools better than public schools? New book says 'no'", Washington Post, Nov 5, 2013; and Christopher and Sarah Lubienski, "The Public Advantage: Why Public Schools Outperform Private Schools", University of Chicago Press, 2014.

[304] U.S. Bureau of Labor Statistics, labor category 25-2031. CPI: U.S. Bureau of Labor Statistics

[305] National Center for Education Statistics

[306] This table is my summary interpretation of the top 5 most relevant myths from David Berliner, Gene V Glass, and Associates, "50 Myths & Lies that Threaten America's Public Schools: The Real Crisis in Education", Teacher's College Press, Columbia

University, New York, 2014. Each myth and lie is considered in detail in this work, including data sources.

[307] Valerie Strauss, "The real effect of teachers union contracts", Washington Post, Oct. 25 2010

[308] Chapter B, "Financial and Human Resources Invested in Education", Education at a Glance, OECD 2011, Chart B1.5.

[309] Richard Rothstein with Karen Hawley Miles, "Where's the Money Gone? Changes in the Level and Composition of Education Spending", Economic Policy Institute, 1995. Spending in nine "typical" districts was itemized, comparing 1967 and 1991.

[310] Chambers, Parrish, & Harr, Special Education Expenditure Project (SEEP) "What are We Spending on Special Education Services in the United States, 1999-2000?", U.S. Department of Education, American Institutes for Research, Contract # ED99CO00091.

[311] See also Idea Money Watch's response to "Something Has Got to Change: Rethinking Special Education", http://ideamoneywatch.com/balancesheet/?p=297

[312] Nathan Levenson, "Boosting the Quality and Efficiency of Special Education", Sep 2012, Foreward by Chester Finn Jr and Michael Petrilli, p 2.

[313] Myth #9 "Teachers are the most important influence in a child's education", from David Berliner, Gene V Glass, and Associates, "50 Myths & Lies that Threaten America's Public Schools: The Real Crisis in Education", Teacher's College Press, Columbia University, New York, 2014

[314] Even weak schools continue to be accredited – see "Is higher-ed loan system like Enron?", Dante Ramos, Boston Globe, June 19, 2015; "Our Universities: The Outrageous Reality", Andrew Delbanco, July 9, 2015; and "Let's Kill All the Accreditors", Richard Vedder, Forbes, 6/15/2015

[315] http://www.capenet.org/facts.html

[316] The U.S. ranked 7th out of 100 countries studied worldwide-- Open Budget Survey, 2012, International Budget Partnership.

[317] See, for example, "Following the Money 2014: How the 50 States Rate in Providing Online Access to Government Spending Data", U.S. PIRG Education Fund, Benjamin Davis and Phineas Baxandall, April 2014.

[318] World Justice Project Rule of Law Index 2015, Chief Research Officer Alejandro Ponce, Washington, DC.

[319] See judicialnominations.org, and "Mitch McConnell Tries to Rig The Courts By Blocking Dozens of Obama Judicial Nominees", Jason Easley, 7/6/2015, www.politicususa.com

[320] Molly Hennessy-Fiske, "Immigration: 445,000 awaiting a court date, which might not come for 4 years", LA Times, May 16, 2015.

[321] https://www.federalregister.gov/uploads/2015/05/Federal-Register-Pages-Published-1936-2014.pdf

[322] http://www.irs.gov/instructions/i1040a/ar03.html

[323] See National Taxpayer Advocate Report to Congress, 2014.

[324] Congressional Budget Office, "The Distribution of Household Income and Federal Taxes, 2011", November 2014.

[325] See, for example, "Who Pays? A Distributional Analysis of the Tax Systems in All 50 States 5th Edition", Institute on Taxation & Economic Policy, January 2015, Washington, DC, www.itep.org.

[326] But one example: 160 million x 20 hours each x 7.25 $/hour (U.S. federal minimum wage) = over 23 Billion $$ spend complying with the IRS, conservatively. Source: National Taxpayer Advocate Annual Report to Congress, Volume 1, 2014.

[327] Office of Management and Budget, "Information Collection Budget of the United States Government", 2014.

[328] Simon Johnson, Daniel Kaufmann, and Pablo Zoido-Lobatón, "Regulatory Discretions and the Unofficial Economy", American Economic Review, May 1998

[329] Worldwide Governance Indicators Project, info.worldbank.org/governance/wgi/index.aspx#home.

[330] Union of Concerned Scientists, "Overwhelming Risk Rethinking Flood Insurance in a World of Rising Seas"

[331] Xun Yao Chen, "Must-know: Important types of farm subsidies in the United States", Market Realist, Oct 10, 2013.

[332] Robert McIntyre et al, "The Sorry State of Corporate Taxes What Fortune 500 Firms Pay (or Don't Pay) in the USA And What they Pay Abroad 2008-2012" Citizens for Tax Justice & the Institute on Taxation and Economic Policy, February 2014.

[333] R. Sharp & U.R. Sumaila, "Subsidies to U.S. Fisheries", Lenfest Ocean Program Research summary of findings, Quantification of U.S. marine fisheries subsidies, North American Journal of Fisheries Management, February 2009.

[334] Sources: Fishing: R. Sharp & U.R. Sumaila, "Subsidies to U.S. Fisheries", Lenfest Ocean Program Research summary of findings,

Quantification of U.S. marine fisheries subsidies, North American Journal of Fisheries Management, February 2009; Corporate Tax: Robert McIntyre et al, "The Sorry State of Corporate Taxes What Fortune 500 Firms Pay (or Don't Pay) in the USA And What they Pay Abroad 2008-2012" Citizens for Tax Justice & the Institute on Taxation and Economic Policy, February 2014; Agricultural: Xun Yao Chen, "Must-know: Important types of farm subsidies in the United States", Market Realist, Oct 10, 2013; Nuclear: Joe Romm, "How much of a subsidy is the Price-Anderson Nuclear Industry Indemnity Act?", thinkprogress.org, Aug 7, 2008; Ethanol: Low: Christopher R. Kittel, "Corn Belt Moonshine The Costs and Benefits of US Ethanol Subsidies", American Boondoggle: Fixing the 2012 Farm Bill, American Enterprise Institute. High: Adenike Adeyeye et al, "Estimating U.S. Government Subsidies to Energy Sources: 2002-2008", Environmental Law Institute, September 2009; Energy Subsidies (including fossil fuel): Low: Figure 21. High: Figure 22.

[335] For example, Fisher, J.C. and R.H. Pry, A Simple Substitution Model of Technological Change, *Technological Forecasting and Social Change*, Vol 3, Pages 75 – 99, 1971

[336] William Nordhaus, "The Climate Casino: Risk, Uncertainty, and Economics for a Warming World", Yale University Press, 2013, Chapter 16 and 18. Using a 4% discount rate NPV calculation, while more complicated than payback period calculations, includes the time value of money, alternative investment options, and risk.

[337] Note, I have not done a comprehensive cost-benefit analysis on this, and it may be more appropriate to offer low interest rate financing, rather than free.

[338] Note, we must be careful not to go too far relative to energy efficiency. In Japan, for instance, energy efficiency standards have increased so much that some projects will not pay back over their lifetime.

[339] http://www.taxpolicycenter.org/taxfacts/displayafact.cfm?Docid=205

[340] Also called a "principal agent" market failure. The landlord pays for the building and energy efficiency equipment – insulation,

windows, air conditioning; but the tenant gets the benefits of the savings if high efficiency equipment is purchased (generally at higher cost). Because the landlord doesn't get any of the savings, he or she won't spend more on higher efficiency products. This also can apply to builder/owner relationships.

[341] Also called the "split incentives" problem, or principal-agent problem. See also "Mind the Gap: Quantifying Principal-Agent Problems in Energy Efficiency, International Energy Agency, 2007; and http://www.rmi.org/Knowledge-Center/Library/2012-05_GuideForLandlordsTenants

[342] Very roughly converting gross output to GDP by dividing by 2 (see "What is gross output by industry and how does it differ from gross domestic product", www.bea.gov)

[343] http://aceee.org/portal/national-policy/international-scorecard

[344] See Vaidyanathan et al, "Overcoming Market Barriers and Using Market Forces to Advance Energy Efficiency", Mar 2013, American Council for an Energy-Efficient Economy. They estimate 20% savings is available. This is reasonable to first order accuracy, and 2.3% of U.S. GDP is spent on electricity alone (EIA). 20% savings is therefore $\sim$.5% GDP.

[345] Tom Randall, "High Cost of Climate Earns Exxon Rare Environmental Win", 3/21/14, Bloomberg

[346] Rebecca Leber, "Republican Senators Finally Admit That Climate Change Is Not A Hoax But they still insist that humans don't have anything to do with it", The New Republic, Jan 21, 2015

[347] See http://www.skepticalscience.com/human-fingerprint-in-global-warming.html for proof of: (a) human emissions are increasing, reducing O_2 levels as we burn fossil fuels; (b) less heat escapes to space consistent with higher levels of greenhouse gas emissions; (c) there are more warm nights, which means that solar only effects cannot explain global warming. See also: Zain Shauk, "Oil CEO: Humans are involved with climate change", Houston Chronicle, May 15, 2013.

[348] "Updated Capital Cost Estimates for Utility Scale Electricity Generating Plants", April 2013, U.S. Energy Information Administration. Estimate is the average for Coal Plant overnight capital cost for plants without CCS, Table 1, in 2012 $.

[349] Metin Celebi, "Coal Plant Retirements and Market Impacts", The Brattle Group, Feb 5, 2014, presented to Wartsila Flexible Power Symposium 2014 – Vail Colorado, slide 5.

[350] See Gary Charbonneau, "Coal", Discussion Sheet Prepared by Peak Oil Task Force, https://bloomington.in.gov/media/media/application/pdf/4719.pdf. EIA coal economist George Warholic states that "EIA's current inability to incorporate updated reserve data into [its] existing database" because of lack of funding

[351] David Rutledge, "Energy supplies and climate policy", May 2012, http://judithcurry.com/2012/05/03/energy-supplies-and-climate-policy-2/

[352] J. David Hughes, "Drill, Baby Drill: Can Unconventional Fuels Usher in a new Era of Energy Abundance?", Feb 2013

[353] For instance, the recent CAFE regulations will reduce gasoline usage in vehicles in the U.S., allowing the industry to sell gasoline for more years (assuming a relatively fixed supply). As extraction costs increase, prices will rise, and these regulations will produce more profit for the industry.

[354] U.S. DOT, vehicle miles traveled per person, U.S. highway statistics

[355] U.S. Energy Information Administration (EIA) for CO_2 emissions; U.S. Census bureau for intercensal population estimates

[356] Mathew Carr, "Rising German Coal Use Imperils European Emissions Deal", June 20, 2014

[357] Greendex 2012: Consumer Choice and the Environment—A Worldwide Tracking Survey, National Geographic, GlobeScan Incorporated. Frequency of using local public transportation to save fuel and reduce pollution is lowest for Americans, at 16% of consumers.

[358] U.S. Bureau of Labor Statistics

[359] Ferald & Jones, "The Future of U.S. Economic Growth", January 2014, Federal Reserve Bank of San Francisco Working Paper Series; Paul Gilding "The Great Disruption: Why the Climate Crisis Will Bring On the End of Shopping and the Birth of a New World", 2012. Naomi Klein "This Changes Everything: Capitalism vs. The Climate", 2014. Richard Heinberg "The End of Growth: Adapting to Our New Economic Reality", 2011; Charles Wessner, "Sustaining Moore's Law and the U.S. Economy", The National Academies, IEEE and the AIP, 2003. Jorgen Randers, "2052: A Global Forecast for the Next Forty Years, A Report to the Club of

Rome Commemorating the 40th Anniversary of *The Limits to Growth*," Chelsea Green Publishing, White River Junction, VT, 2012

[360] 2015 Munich Re, NatCatSERVICE, Property Claim Services (PCS), a Verisk Analytics business, as of June 2015. Graphs accessed from Insurance Information Institute, www.iii.org on July 30, 2015.

[361] Ackerman et al, "The Cost of Climate Change: What We'll Pay if Global Warming Continues Unchecked", May 2008, NRDC. "The Cost of Delaying Action to Stem Climate Change", July 2014, Executive Office of the President of the United States

[362] Robert Ferris, "Honeybees are dying, and scientists still don't know why", CNBC, May 14, 2015

[363] Portnov et al, "Offshore permafrost decay and massive seabed methane escape in water depths >20m at the South Kara Sea shelf", Geophysical Research Letter, V40, #15, 16 Aug 2013

[364] Rahmstorf et al, "Exceptional twentieth-century slowdown in Atlantic Ocean overturning circulation", Nature Climate Change, Vol5, May 2015; Chris Mooney, "Why some scientists are worried about a surprisingly cold 'blob' in the North Atlantic Ocean", Washington Post, Sep 24, 2015

[365] Brienen et al, "Long-term decline of the Amazon carbon sink", Nature, V519, pp 344-8, 19 Mar 2015

[366] Karl et al, eds. "Global Climate Change Impacts in the United States", United States Global Change Research Program (USGCRP), 2009, Cambridge University Press, New York, New York.

[367] AR5, IPCC.

[368] One tonne = a metric ton = 1000 kg.

[369] EIA data per Figure 11 for 1949-2014; Oakridge National Laboratory Carbon Dioxide Information Analysis Center, T.A. Boden, G. Marland, and R.J. Andres Global Regional and National Fossil Fuel CO_2 Emissions DOI 10.3334/CDIAC/00001_V2010 prior to 1949.

[370] See www.epa.gov/superfund; www.hanford.gov/page.cfm/HanfordCleanup; John Berg, "The cleanup of Boston Harbor was surprisingly triumphant", Aug 1 2004, Commonwealth Politics, Ideas, & Civic Life in Massachusetts; ozonewatch.gsfc.nasa.gov; www.epa.gov/acidrain/

[371] Stephen Pacala and Robert Socolow, "Stabilization Wedges: Solving the Climate Problem for the Next 50 Years with Current Technologies", Science, August 13, 2004, Vol 305, #5686, pp 968-972. DOI:10.1126/science.1100103

[372] EPA, http://www3.epa.gov/epawaste/nonhaz/municipal/pubs/2012_msw_fs.pdf

[373] See Craighill and Powell, "Lifecycle assessment and economic evaluation of recycling: a case study", V17, 1996, Resources, Conservation and Recycling. This is one of hundreds of similar studies. Note that it is critical to include all costs and benefits of recycling on a lifecycle basis, as the primary benefit of recycling is avoidance of virgin material production costs.

[374] Michael Munger, "Recycling: Can It Be Wrong, When It Feels So Right?", www.cato-unbound.org

[375] Jennifer Nash and Christopher Bosso, "Extended Producer Responsibility in the United States.", Journal of Ecology, 17.2 (2013), pp 175-85

[376] U.S. EPA

[377] Dana Gunders, "Wasted: How America is Losing Up to 40 Percent of Its Food from Farm to Fork to Landfill", Aug 2012, NRDC

[378] Laura Wellesley, Catherine Happer, and Antony Froggart, "Changing Climate, Changing Diets: Pathways to Lower Meat Consumption", Chatham House Report, Nov 2015; and Stehfest et al, "Climate benefits of changing diet", Climate Change, 2009, V95, pp 83-102.

[379] http://www3.epa.gov/climatechange/ghgemissions/sources/agriculture.html

[380] "66 Ways to Absorb Carbon and Improve the Earth's Reflectivity, From Reasonable Options to Mad Scientist Solutions", Risto Isomaki, Into Publishing, 2009, Ways # 27, 32, 35, and 42.

[381] A lot of instruments are needed for large areas over large periods of time, and experiments are difficult to have a proper control

[382] For instance, Risto Isomaki, in "66 Ways to Absorb Carbon and Improve the Earth's Reflectivity, From Reasonable Options to Mad Scientist Solutions", Into Publishing, 2009, Way #20.

[383] Eilperin, Juliet. "Great Barrier Reef has lost half its corals since 1985, new study says". The Washington Post. Retrieved 1 October 2012

[384] "66 Ways to Absorb Carbon and Improve the Earth's Reflectivity, From Reasonable Options to Mad Scientist Solutions", Risto Isomaki, Into Publishing, 2009, Way #20.

[385] If Encyclopedia Britannica is correct, that limestone represents 1% of the earth's total volume, then 12,000 million tons of limestone used annually would last 250 million years, well past the time our fossil fuel resources are exhausted.

[386] Perhaps using sodium silicate as an adhesive if necessary

[387] http://costs.infomine.com/costdatacenter/miningcostmodel.aspx

[388] Crushed stone, mostly limestone, is sold in the U.S. today for an average of ~10 $/tonne. (mineral.usgs.gov)

[389] David Keller, Ellias Feng, & Andreas Oschlies, "Potential climate engineering effectiveness and side effects during a high carbon dioxide-emission scenario", Nature Communications, Vol 5, #3304, Feb 2014

[390] See also L.D.D. Harvey, "Mitigating the atmospheric CO2 increase and ocean acidification by adding limestone powder to upwelling regions", Journal of Geophysical Research, V113, C04028, doi:10.1029/2007JC004373, 2008.

[391] www.rustletheleaf.com/10things.html

[392] www3.epa.gov/region09/newsevents/images/40things-poster.pdf

[393] "Never doubt that a small group of thoughtful, committed citizens can change the world; indeed, it's the only thing that ever has.", attributed to Margaret Mead, in Nancy C. Lutkehaus,"*Margaret Mead: The Making of an American Icon*" (Princeton, NJ: Princeton University Press, 2008), p. 261

[394] Inventory of U.S. Greenhouse Gas Emissions and Sinks: 1990-2010. April 25, 2012, EPA, EPA 430-R-12-001.

[395] "Energy Use in the U.S. Steel Industry: An Historical Perspective and Future Opportunities", Dr. John Stubbles, Sept 2000, for U.S. Dept. of Energy Office of Industrial Technologies, Figure 6 on top. Theoretical limit from p 3 of "The Carbon and Energy Intensity of Manufacturing", Timothy G. Gutowski, 40th CIRP International Manufacturing Systems Seminar at Liverpool University, 30 May to 1 June 2007.

[396] "A Comparison of Iron and Steel Production Energy Use and Energy Intensity in China and the U.S.", Hasanbeigi et al, June 2011, LBNL-4836E., Figure 2.

[397] www.wikipedia.org. "Blast Furnace", "Open Hearth Furnace", "Electric Arc Furnace"

[398] www.recycle-steel.org, Steel Recycling Institute steel recycling rates announcement.

[399] The percentage of EAF vs. other processes is a key driver in the difference between China and U.S. steel industry emissions. See "A Comparison of Iron and Steel Production Energy Use and Energy Intensity in China and the U.S.", Hasanbeigi et al, June 2011, LBNL-4836E.

[400] See, for example, "EAF and/or BF/BOF Which route is best for Europe?", Marcel Genet and Laplace Conseil, Laplace Consel Consulting, 2012, for further details on these issues.

[401] "Global Warming Impact on the Cement and Aggregates Industries", Joseph Davidovits, in World Resource Review, V6, #2, pp 263-278, 1994, p 3.

[402] UWM Center for By-Products Utilization. "Global Warming and Cement-Based Materials", Tarun Naik and Rakesh Kumar, June 2010. Corrosion inhibitors can potentially compensate for the disadvantages.

[403] One example is Novacem, which uses MgO

[404] www.wikipedia.org, "Geopolymeric Cement"

[405] www.wikipedia.org, "Portland Cement"

[406] This is less common in the U.S., but does occur overseas. The flaring reduces the methane to CO_2 to minimize its environmental impact.

[407] U.S. Energy Information Administration (EIA)- Annual Energy Outlook (AEO) 2012.

[408] See the recent furnace and boiler rulemaking, http://www1.eere.energy.gov/buildings/appliance_standards/rulemakings_and_notices.html

[409] See the Department of Energy's Furnace and Boiler rulemakings website for further details. To avoid grandfathering issues as occurred with the EPA clean air act and coal generation, funds for retrofits will likely be needed

[410] http://www.epa.gov/climatechange/ghgemissions/gases/n2o.html

[411] "Fertilizers and Climate Change: Enhancing Agricultural Productivity and Reducing Emissions", 24 July 2009, www.fertilizer.org

[412] See http://water.epa.gov/polwaste/nps/outreach/point1.cfm

[413] http://www.newairplane.com/787/design_highlights/#/exceptional-value/lower-fuel-consumption

[414] "Evolution of the Airliner", Ray Whitford, 2007, p 117

[415] U.S. Census of agriculture

[416] Isolation of Succinivibrionaceae Implicated in Low Methane Emissions from Tammar Wallabies P. B. Pope, Science 333, 646 (2011)

[417] http://mercatus.org/publication/assessing-department-energy-loan-guarantee-program

[418] Richard Baldwin & Frederic Robert-Nicoud, "Entry and Asymmetric Lobbying: Why Governments Pick Losers", Feb 2007.

[419] Jenkins, Swezey, and Barofsky, "Where Good Technologies Come From: Case Studies in American Innovation", Dec 2010, Breakthrough Institute.

[420] ROI = return on investment

[421] See, for instance, http://www.stage-gate.com/resources_stage-gate_full.php

[422] See Robert Cooper, "Winning at New Products: Creating Value Through Innovation", 4th Ed, Basic Books, New York, 2011.

[423] https://www1.eere.energy.gov/manufacturing/financial/pdfs/itp_stage_gate_overview.pdf

[424] William Nordhaus, "The Climate Casino: Risk, Uncertainty, and Economics for a Warming World", Yale University Press, 2013, Chapter 21.

[425] "Because the issue of climate change is global, it must become a truly global concern as well. All developed and developing economies, particularly India and China, can make significant contributions in dealing with the matter. It would be unrealistic and counterproductive to expect the U.S. to carry burdens which are more appropriately shared by all." p 35, 2008 GOP Platform, www.gop.com

[426] "China has replaced America as the world's largest emitter of greenhouse gases. …We need a global response to climate change that includes binding and enforceable commitments to reducing emissions, especially for those that pollute the most: the United States, China, India, the European Union, and Russia". 2008 Democratic   platform,

http://www.presidency.ucsb.edu/ws/?pid=78283 See also the 2012 Democratic Platform at www.democrats.org

[427] Simple answers to questions a-d are: yes, yes, 1700, and yes

[428] William Nordhaus, "The Climate Casino: Risk, Uncertainty, and Economics for a Warming World", Yale University Press, 2013, Chapter 12.

[429] Meadows, Randers, & Meadows, "The Limits to Growth," Universe Books, New York, 1972. Meadows, Randers, & Meadows, "Limits to Growth: The 30-Year Update," Chelsea Green Publishing Company, White River Junction VT, 2004. Jorgen Randers, "2052: A Global Forecast for the Next Forty Years, A Report to the Club of Rome Commemorating the 40th Anniversary of *The Limits to Growth*," Chelsea Green Publishing, White River Junction, VT, 2012.

[430] "Limits to Growth: The 30-Year Update," Chelsea Green Publishing Company, White River Junction VT, 2004, Chapter 6. Note, this is the view of the authors of "Limits to Growth", not my own, as described in later chapters. Reprinted with permission

[431] Paul Gilding "The Great Disruption: Why the Climate Crisis Will Bring On the End of Shopping and the Birth of a New World", 2012. Naomi Klein "This Changes Everything: Capitalism vs. The Climate", 2014. Richard Heinberg "The End of Growth: Adapting to Our New Economic Reality", 2011.

[432] Gordon Moore, April 13, 2005. http://en.wikipedia.org/wiki/Moore's_law

[433] www.fao.org, summary tables of fishery statistics

[434] "Fertilizers, Climate Change and Enhancing Agricultural Productivity Sustainability," p 1, International Fertilizer Association, 2009. www.fertilizer.org

[435] James Hansen's "Storm of my Grandchildren: The Truth about the Coming Climate Catastrophe and Our Last Chance to Save Humanity", 2009, Figure 16, showing the concept of the airborne fraction whereby 44% of the CO_2 we emit into the air is disappearing into sinks—the ocean, forests, soils.

[436] See Raymond Pierrehumbert, "Infrared radiation and planetary temperature", January 2011, Physics Today, p 33-38, for a much more detailed explanation and proof of this effect. Saturation effect fallacies, and why CO_2 is more critical than water vapor, are also discussed.

[437] See http://earthobservatory.nasa.gov/Features/GlobalWarming/page2.php

[438] U.S. Geological Survey Earth Science Information Services. http://soundwaves.usgs.gov/2012/06/, accessed 09/2014.

[439] G.R. Dickens, "Down the Rabbit Hole: toward appropriate discussion of methane release from gas hydrate systems during the Paleocene-Eocene thermal maximum and other past hydrothermal events", Climate of the Past, V7, p 831-46, 2011. This is but one of many controversial papers on the subject.

[440] IPCC AR5, Box 6.3. There is significant uncertainty associated with N_2 availability, drought, resistance to drought, and other limits that may affect CO_2 fertilization; however, the CO_2 fertilization effect makes it less likely that Amazonian or Boreal forest die-off could occur.

[441] Once additional growth occurs, when the trees die they release their carbon, with births balancing out this release thereafter.

[442] https://nsidc.org/cryosphere/frozenground/methane.html

[443] See www.wikipedia.org/weathering

[444] David Murphy and Charles Hall, "Year in Review—EROI or Energy Return on Energy Invested", Annals of the New York Academy of Sciences, Volume 1185, 2010, pp 102-118, New York Academy of Sciences. Note, many of these values are controversial.

[445] David Murphy, public domain, https://en.wikipedia.org/wiki/File:Net_energy_cliff.gif

[446] There have been some recent studies that claim much higher EROEI for fracking natural gas, which used one of the best fracked wells in central Marcellus Shale as being representative of all fracked wells. Well quality appears to drop away farther from the center of leading formations, leading to the lower figure reported here.

[447] Note, these 2010 EROEI figures presented are static and do not take into account technical innovation over time. For example, in 2010, EROEI for solar was ~7. But prices for solar have dropped by more than 30% since, so the EROEI for solar is actually closer to 10 today. Similar arguments apply to most of the sources in this figure, so this graph may be pessimistic by ~20-30%. On the other hand, these EROEI figures may not go far enough relative to included costs (the nuclear EROEI discounts the cost of nuclear

waste and public liability costs; ethanol costs may not count the fertilizer needed to restore the soil fertility the corn uses, and fossil fuel estimates do not include potential impacts or costs of CO_2 emissions, etc.) and may therefore be overstated. EROEI measurements are controversial, and depend on one's viewpoint.

[448] Charles Hall, Stephen Balogh, and David Murphy, "What is the Minimum EROI that a Sustainable Society Must Have?", Energies Journal, Volume 2, #25-47, 2009.

[449] U.S. Energy Information Administration, Annual Energy Outlook April 2015, DOE/EIA-0383 (2015), Table A11.

[450] David Rutledge, "Estimating long-term world coal production with logit and probit transforms", International Journal of Coal Geology, July 2010.

[451] See Gary Charbonneau, "Coal", Discussion Sheet Prepared by Peak Oil Task Force, https://bloomington.in.gov/media/media/application/pdf/4719.pdf. EIA coal economist George Warholic states that "EIA's current inability to incorporate updated reserve data into [its] existing database" because of lack of funding

[452] David Rutledge, "Energy supplies and climate policy", May 2012, http://judithcurry.com/2012/05/03/energy-supplies-and-climate-policy-2////

About the Author

 Graham Stevens is a consultant with over 25 years of experience in consulting, manufacturing, reliability, and R&D. He specializes in operations enhancement and bringing a broad range of products from the laboratory into production.

Project work has included detailed factory due diligence (hundreds of factories), process engineering, photovoltaic technology (all types), energy efficiency, financial modeling, renewable technologies (wind, biomass, etc.), market strategy (market entry, benchmarking, market penetration modeling, job assessments, business model/value chain determination, market competitiveness), road-mapping, supply chain assessment, make/buy decisions, renewables integration, project management, reverse engineering, operations enhancement, and clean tech investment. He builds step-by-step cost models to predict high volume production costs from napkin sketches, teaches Design for Manufacture and Assembly, and focuses on cost reduction and lean engineering.

He can be found at www.costreductioninc.com.